IN PURSUIT

A Pilot's Guide to Online Air Combat

JOHAN KYLANDER

PILOT PRESS

For Barry "Carrot" Hayes

Published by Pilot Press, Stockholm, Sweden
ISBN 91-976077-0-3

Typeset in Garamond, 12p

PILOT PRESS
Päronvägen 23, 146 48 Tullinge, Sweden. Email: jkyl@comhem.se

JUNE 12: TRIPLE TRIP

I step into my personal P40-B at Cambrai, strap in, slide the canopy back and go through the pre-launch routine: tailwheel locked, trim wheel fully back, RPM set at maximum, listen and scan for intruders. None around. It's 13:45 local time and the hound dogs are too lazy to even yap in the blistering French afternoon. I fire up the engine and slam the throttle full forward, clearing the south hangers with feet to spare. I ess and quarter-roll all the way up to 4 km where I reduce to normal pitch and cruise east over the battlefield, still quarter-rolling and weaving gently. I check my six every 2 seconds.

There's nothing to be seen at the Meuse apart from a grand vista at 4,5 km. No enemy formations, not even a single recon plane. Oh well. I cut to fine pitch, anticipating a boring lonely wait over Bertrix-Acremont. I'm doing about 340 IAS as I come in from southwest of the field. Hey, there's a bogey diving in west at about 4 km – probably friendly. I lose him in the clouds as I swing left to take up orbit. 2 seconds later I spot a 109 arcing in on my six from the west, co-alt. Well well, who'd a thunk! Someone else must have been here to stoke the waffles not long ago. I go to max revs and waver a second while assessing his energy. Co-E, or nearly so. Not yet in guns range, not even close. I swing into him to left and cross over for a lead turn to right. We're canopy to canopy by then, so I lag him a bit and let him do the hauling. We're in a rolling scissor for a little while and then he goes vertical, sort of. Damn those Fokkers are fast, but not fast enough to draw out of .50 range. I reach out and touch him, drawing a fine trail of fluids.

The fight goes downward to about 3 km. I get a few snapshots, some of them connect briefly. He struggles and hauls for all he's worth, blowing energy in chunks. I slowly reel him in, stitch him a little more. He tries to go vertical again and right there I have him. I give him a long nice burst and watch him disappear in a million pieces – just as I catch a glimpse of another 109 nearby. This guy appears to have engaged from below and is quite out of steam. I reel him in too and punch his lights out. We're now uncomfortably close to the field and it's time to say goodbye! I scoot out north and disengage entirely to the east before starting a slow climb back to altitude. I visit Wiltz – no trade – and debate whether to turn back. No, Verviers-Aachen it is.

Verviers looks empty as does the surrounding lanes to west and northwest. I carry on to Aachen. Upon arrival, I spy a single-engine job rotate and head out westerly. I circle astern, let down and roll into a gentle dive. 109. He's in a hurry. He's not checking six. I hold the shot for too long and only strike him half-heartedly at about 70 meters range. I see my MG splash both his wings then I have to duck violently to avoid the collision. Red-out. I come to and spiral, tally the bandit at my long low six in a flat circle on the deck, obviously distressed. I go in again, twice, but he's awake now. We're pretty low, much too close to Aachen for comfort. I rope-a-dope him. The bastard refuses to stall out but lifts his nose for a head-on shot. I'm as surprised as I'm disgusted. I see my hits splash all over his engine and forward fuselage. I have the energy to decide though against this Emil, and disengage. He follows.

I drag him along for a while, heading south. I've had enough near Eupen and chandelle around to have another go at him. Again he comes head-on. Oh come on! I press the tit bitterly and give him a long good burst, skidding slightly. Die! And so he does. I've had enough of the likes of him and return to base, booking three confirmed kills.

BMBM, 56th FG(v)

INTRODUCTION

Greetings to all Virtual Fighter Pilots and those "want-to-be" Pilots. This Work-of-Art "In Pursuit" by Johan Kylander is a MUST READ.

In Pursuit is a thorough anthology of all facets of the life of a fighter pilot. In Pursuit covers, in detail, all of the positives and negatives, the things to do and the things not to do in order to become a "Top Gun" as a fighter pilot. Johan Kylander has given a lot of time and thought into developing a complete and easy to read compilation of all phases of tactical flying.

When I was assigned to the 56th Fighter Group, 61st Fighter Squadron. I was introduced to Flight Leader "Mike" Gladych (PAF) and assigned to him as his wingman. Many things he taught me still remain in memory. One of the most important was this: if you know your machine better then your opponent knows his, you can always defeat him. The next best was this: fly your position (on his wing) where I tell you to and I will always bring you home. (Correct positioning of all the elements in the formation is the best protection you can give to each other). I flew fiftyseven missions over Europe during WW II and I feel like I remember a little of what I was taught. I've learned a lot of things from reading Johan Kylander's book "In Pursuit" that I wish I had known way back then. I would have been a much more successful fighter pilot.

Russell S. Kyler
Captain USAAF (1945)

PREFACE

Walter Mitty exists. Every guy or gal who has ever sat down in front of the computer to wreak havoc in the virtual skies of Air Warrior, WarBirds, Aces High or World War II Online, embodies the spirit of Walter, that paragon of daydreaming indulgence in personalities and situations that he never had the opportunity of experiencing himself. Much like Walter, we virtual pilots suspend reality and engross ourselves in the fantasy of being Sierra Hotel fighter pilots, flying and fighting over war-torn Europe, seat-of-the-pants style. Some of us do. Others, to the utter scorn of the historically versed grognards, are just as happy whiling away a few hours without a thought of role-playing, yanking and banking to get an adrenaline rush in what might just as well be another video game. Either is fine. We're all warriors.

Yet, whether one seeks the thrill of the moment or strives to recreate an environment approximating history, one trait is common: the will to succeed and to excel. And that is what this book is about – to help neophyte and intermediate virtual pilots surpass the thresholds to acedom.

Learning how to play a computer game should be an easy matter. Most games ARE easy to master once you've figured out what makes them tick, but the massively multiplayer online air combat game is a striking exception because it isn't predictable. There are simply too many factors to consider at any turn: beyond certain automated functions the action is totally unscripted, unregulated and unpredictable because every action or piece of the aerial environment is predicated by another human being - and it all happens in real time. Fact is, the difference between fiction and reality has a tendency to narrow to such an extent that some would argue that the game IS reality. In certain respects that is entirely true.

Aircraft performance, tactics, psychology and physics apply to this virtual world on the same principles as in the real world, to extents varying with the creators' vision and programming acumen. When it comes to engagements, that which works – or doesn't work – in real life, also works or doesn't work in the game. Thus, knowing a thing or two about these matters significantly increases the virtual pilot's likelihood of excelling. Other games allow its players "crutches" such as external views or the use of shortcuts to attain their goals. There are no shortcuts or power-ups in the virtual sky. You can't fake acedom.

Air combat seems so easy... all you need to do is to point your aircraft at another and gun him down, right? Not! It is no coincidence that aerial combat skills are most difficult to acquire - we're talking four dimensions here, where the input is constantly changing. He who is not able to arrive at the correct solution at every juncture will soon find himself involuntarily decorating the landscape.

Some are born fighter pilots. If you fail to return to base on a regular basis, take it as a hint that you're not one of them. You need to practice and practice and practice yet more to reduce your errors to a survivable level, and even so there's still the odd chance that you'll meet the Red Baron on your next mission – so be prepared!

I decided to write this book more or less on a whim. Over the years of online combat I've authored countless pilot FAQ's, training manuals, squadron directives and forum posts dealing with this or that tactical problem. I'm not a real pilot – I have no formal training other than a few hours of introduction flight in civilian craft, but I do however have something like 8000 hours of virtual combat to lean on and a comfortable factual background gleaned from such outstanding works as "Boots" Blesse's "No Guts, No Glory", Robert "Mouse" Shaw's "Fighter Combat: Tactics and Manoeuvring", the "Bible" of all sim pilots, and others too many to mention. While the "Bible" is certainly a worthwhile read and in nearly all aspects entirely relevant to our environment, it is nevertheless a tad dry and not quite as accessible to the beginner virtual pilot as one might want. Furthermore, Shaw's work is largely concentrated on the individual aspects of air combat. Some key missing features which I hope to add here includes the psychology of online fighting and a look inside the opponents' brains – the understanding of which are crucial for success. Ergo, this attempt at writing a practical guide to air combat.

Enjoy.

Johan "BMBM" Kylander
Stockholm, May 2005

FOREWORD TO THE PRINTED EDITION

This book was originally published online and will remain available for free at http://tinyurl.com/ehstj for the duration. I am deeply indebted to Mr James Tolve of Boulder, Colorado for his selfless contribution to making this book available in printed form. Without his encouragement and help this project would still be on the back burner. Jim is truly the best wingman a guy can wish for. I would also like to thank Mr Edmund Gregr or Plzen, Czech Republic, for his unstinting support and outstanding friendship in this and other endeavours. May your camels never lack water.

In Pursuit in the printed edition is somewhat different than the online work: much additional text has been incorporated and the illustrations are brand spanking new and improved. Where I have added to the original text it has been for the sake of clarity and ease of understanding of the sometimes very complicated chain of minute events that is online air combat.

I would also like to thank the community for the very rewarding feedback I have been given since making this book available. Verily it has been and will remain my most cherished reward: to know that virtual pilots everywhere have benefited in some way from my contribution.

Check six,
Johan "BMBM" Kylander
Stockholm, October 2006

CONTENTS

PART I: THE ENVIRONMENT

Muster was called at 3:30 PM. Due to the fact that the rest of my squadron had not yet returned from R&R, I was temporarily assigned to an ad-hoc strike force from REIMS to Buzancy. Inexperienced and/or poor leadership and bad judgement are often the bane of any military endeavour - and this was no exception. Our flight leader flew us in to Buzancy at 1.5 Km. A 109E went past us high and to the north about 15 miles west of the target area, which I reported to my flight leader. The flight was instructed to ignore the enemy aircraft - but it would have been more productive to instruct the 109 to ignore us!

Guy "Ghastly" Skaggs, 56th FG (v)

CHAPTER 1: THE WORLD

The online "world" is a persistent universe, an open arena, populated by fun-seekers and with whatever structures and automated features the game designers provide it with. It's a 24/7 battlefield where combat is sought and generally found within a few minutes of entering the world ("spawning in"). For the purposes of this book, the bulk of references to the world, its vehicles, features and inhabitants, are drawn from the Massively Multiplayer Online Role Playing Game (MMORPG) World War II Online (WWIIOL). The game attempts to recreate combat in Europe 1940-43, at the tactical and, to some extent, the strategic level. WWIIOL is a significantly different environment from other games inasmuch as it incorporates all aspects of the virtual battlefield – it joins air combat with infantry and tanks, surface and submarine vehicles, supply, interdiction and attrition. Yet, for the purpose of air combat, action in WWIIOL is much the same as is found in other games such as WarBirds and Aces High: everyone is out to kill someone, or as the more politically correct would have it, to spend quality time interacting with other people, online.

When entering the world to strap into your aircraft of choice, you are presented with a map view. In WarBirds, Aces High et al, the world is divided between two or more sides, up to a default maximum of four, each having access to a multitude of bases or spawn points. From there on it's up to you, the player, to decide where to spawn and with what equipment, which type of mission to fly etc. Typically, there is a frontline of sorts where you can be reasonably certain of finding action within minutes or even seconds of launching into the world, and there is also always a rear area where action is possible but less likely. In any event, there are seldom places where you can be totally certain of NOT finding someone intent on your destruction.

MMORPG are different from "standard" online games primarily in the sheer numbers of opponents. A "standard" online game such as IL-2 might offer a semi-scripted scenario where up to 16, 24 or at best 64 players can partake. WarBirds and WWIIOL on the other hand, can easily accommodate hundreds, thousands of players simultaneously (although the maximum viewable players in any one local area is limited to 128 – a legacy of internet communication protocol). This naturally means that no mission is ever the same as the other and that you're less likely to encounter any particular enemy twice under similar circumstances. In this world, nothing is scripted. Whatever action is encountered is the result of the players' premeditated choice. In other words, the player is master of the situation. This is important to remember later on, that *the only thing limiting the player's engagement circumstances is the player himself.*

When you enter the game world at your location of choice, you're already sitting at the controls of an WWII-era aircraft. Your field of view is that of the pilot's, it's a first-person perspective closely resembling that of what a real pilot would see. The notable difference is lack of depth and lack of peripheral vision, something that is a critical issue when considering what we call situational awareness. A real pilot would probably liken the experience to flying with blinkers – the virtual pilot is forced to turn his attention in a dedicated manner to this or that direction, he has little sense of what goes on to his sides or rear unless he actually looks specifically in those directions. That's bad news for formation flying although it certainly doesn't make formation flying an exercise in futility. It just makes it a wee bit more challenging.

The first MMORPG worlds were rather bleak – they offered a perpetually CAVU[1] sky and a little less than totally flat terrain until recently when computing power became more potent and accessible. That was fine at the time, particularly since people were happy to do battle at all on the massive scale and since variations in terrain and weather would be an obstruction to the core experience – the pure fight. I remember the polygon worlds of early flight sims – we were delighted by volcanoes 12,000 feet high, flat islands with but a single 1500 meter airstrip and way south on hidden spots, a T-Rex or Godzilla on Gilligan's Island. Lately however, the discerning gamer has come to appreciate and crave more varied experience. Rolling terrain, realistic structures and city layouts, realistic vegetation, weather effects, lighting and volumetric clouds enhance the gaming experience not only from a visual point of view but from an engagement point of view too. In short, pilots appreciate realism and have a sense for beauty in the environment. Imagine that!

This trend is highly likely to continue. In the future, we'll see yet more realistic weather, yet higher resolution of natural features and yet better viewing technology (3D glasses anyone?) to raise the immersion factor yet further. The drawback, of course, is the increasing demands made on the player's hardware (and consequently his budget), which makes the pastime not only time-consuming but frightfully

[1] Ceiling and visibility unlimited

expensive too. Even today the immersion factor is great enough to cause the player to burst out with whoops of joy over a particularly nice sunset or terrain feature, not to mention the pounding sensation of closing on an blissful enemy or the sweating palms resulting from extended hard combat. Immersion is not merely a result of visual and sound effects but of what the player himself contributes in the form of behaviour – most readily seen in formation flying, voice communications and adherence to engagement principles. I'd go so far as to say that the game's appearance means far less in terms of immersion than the level of roleplaying exhibited by the player – if you can suspend disbelief, immersion is bottomless. When we fly and fight, we truly ARE fighter pilots.

CHAPTER 2: THE PLAYERS

In MMORPGs such as WarBirds and WWIIOL players are truly international. While the games are produced and first released in the U.S. they have gathered a strong following all over the world. The main nations besides the U.S. are, in no particular order, Canada, Australia, U.K., Germany, France and Italy. Other nations with relatively strong online communities include Finland, Sweden, Spain and Brazil. Although the player base is mainly (85-90%) rooted in North America, the distribution does provide an 24/7/365 alternate reality.

Regardless of origin pilots can be said to fall into either of two main categories: the casual and the hard-core. This is not indisputably so, to the contrary: depending on your mood, you might be hard-core in one session and casual in another. You may even be a hard-core casual pilot, or a casual hard-core pilot. Still, for the sake of argument, pilots are usually geared either toward a behaviour which is predicated by historical aspects, or toward one that is entirely centred on the fighting experience. I will call them Type A Grognards and Type B Funseekers. The Grognard has a deep interest in the era and has probably spent half a lifetime reading up on the history and particulars of WWII. He knows the difference between MG-FF and MG-151, he can recite orders of battle from obscure engagements over the Owen Stanleys and he knows perfectly well what discerning traits one or the other machine has at varying airspeeds. The Grognard is usually well versed in ACM and formation tactics, and can almost be trusted to immerse himself in the fighter pilot fantasy to the point of wearing silk scarves and goggles. In short, he's a diehard realism freak. The Grognard prepares himself with maps of the arena, he joins up with wingmen and claws for altitude. He abhors "dying", or at any rate does his very best to avoid it.

The Funseeker on the other hand doesn't really bother with specifics: all he wants is a good fight with as little delay as possible. The Funseeker habitually takes on any odds at any time under any circumstance. He usually takes off in the frontline and makes a beeline to the action. If he's shot on the runway, he'll most likely replane and attempt it again no matter what the opposition. The Funseeker is generally a loner. He may know that teamwork improves his survival chances but is at pains to hook up with someone other than by accident. While the Funseeker certainly doesn't enjoy "dying", his flying style is rarely affected by it – although, involuntarily, his anger and frustration over having been whacked might.

Still, these typologies aren't absolute. The great leveller is experience – there are as many inexperienced Grognards as hotshot Funseekers. The game itself also determines the distribution of type A and B pilots – WarBirds for instance has a large cadre of experienced pilots whereas the ace vs. rookie ratio in WWIIOL is something like 1 to 9 or greater, mainly because the WWIIOL player base consists largely of people with tank-infantry background. When these guys go aloft they do so primarily with a view to affect the ground-war situation, which means that they fly low and concentrate on whacking enemy ground units. Needless to say, these "pilots" fall like flies when confronted with anyone having a few years of air-to-air experience.

If you put two pilots of similar experience and capability in two similarly performing aircraft, the one who makes the least mistakes – or he who has the most luck – will win. As luck has it however, no two pilots are hardly ever so closely matched. If you on the other hand put a brilliant pilot in, say, a Brewster Buffalo, and a complete beginner in a Focke-Wulf D-9, chances are that the expert pilot will send the beginner down in flames in very short order. Experience is the deciding matter, not the plane. This was painfully experienced by real aviators in The Great Marianas Turkey Shoot, in the opening days of Operation Barbarossa and in the last year of the European War on the Western front. Experienced, well-trained and highly motivated pilots will sweep the skies every time or at least for as long as they aren't outnumbered twenty to one. Online, it is very much the same: experience is at a premium, and it is dearly purchased through what at times seems to be a near infinite amount of virtual deaths.

Experience will tell you where to launch, where to patrol, how to execute your mission, when to engage and when not to. Experience tells you whether the bogey is friendly or enemy, whether he's a bomber or fighter, whether he's heading toward you or away from you, whether he's seen you or not. Experience gives you a subconscious scanning routine, the ability to format, the tingle along the spine that tells you to break hard away, the ability to check six while you're dogfighting, the keen sense for energy. Experience tells you where your exit window is, when to disengage, how to fox an enemy, the difference between too soon and too late. It is all that you need, and everything the beginner wants.

It's a cruel war out there. Some of your opponents have logged thousands of hours and engaged the enemy tens of thousands times, whereas quite a few have hardly even scored their first kill yet. It is very much like in the real war, but here the veterans are immeasurably more experienced than ever the greatest Experten – except in the discipline of staying alive that is. Until you become experienced, and that may take months, years even, read as much as you can, study acknowledged veterans in the air, and fly as if your life depended on it. You are fortunate in having recourse to a lot of training material and accumulated wisdom that the real pilots didn't – some were sent up in the air with less than 40 hours of tuition, and with no more combat instruction than Dicta Boelcke[1].

---

[1] By the summer of 1916, Oswald Boelcke had become Germany's top fighter pilot. Feldflugchef Colonel Thomsen of the German High Command urged Boelcke to draw up a summary of principles that should govern every air fight. His list of 'rules' for success is often referred to as the 'Dicta Boelcke.'

In the beginning you need to take baby steps and concentrate on a single thing: how to survive. Let me rephrase that: first you must learn to live, not how to kill. As a new pilot you should forget entirely about (dog)fighting, however tempting it may seem to indulge in it. Spend your first month slavishly following other pilots around and STAY with them throughout a sortie. Firing your guns is secondary, survival and not losing your self-chosen leader is primary.

The big question is whom to follow. Choose the ones who climb to at least 3-4 km (9-12,000 ft), the ones who seem to know what they're doing. Forget about the loudmouthed crowd, the weedclippers and the impatient. When you grow confident enough to go on the offensive you should attack only such enemy as are low, slow and non-manoeuvring. These fellows are invariably found enroute to particularly active map areas. Kill them from above and behind when they're napping, with a single devastating high-speed attack.

Immediately break off after an attack, whether successful or not, and regain your advantages and situational awareness. If surprise is lost, disengage entirely: it's hard enough to kill a guy who's sleeping at the stick, killing someone who manoeuvres to avoid you is immeasurably more difficult and quite beyond most raw beginners. And wading right into the furball as a newbie is about the worst thing you can do. Without proper understanding of energy, ACM and combat psychology, and without situational awareness to handle even ONE bandit, how can you hope to prevail? If you perchance return to base it is mainly attributable to dumb luck, numerical superiority and the gracious intervention of friendlies in saving your posterior.

As a rookie pilot you will find advice a dime a dozen, or maybe not at all depending on your co-flyers and community. Hopefully you will pick up some nuggets of wisdom here and put them to use – if so, my ambition has been rewarded.

While I consider myself an accomplished pilot, I am nevertheless just as prone to suffer disaster as anyone else is out there, and perhaps that humility is my saving grace. An example: one fair day I wended my way out over the front in a Dewoitine D.520. I was not far into "indian country", at somewhat below the aircraft's best performance altitude (5500 m), and had been diligently observing the sky ever since

---

1. Try to secure advantages before attacking. If possible keep the sun behind you.
2. Always carry through an attack when you have started it.
3. Fire only at close range and only when your opponent is properly in your sights.
4. Always keep your eye on your opponent, and never let yourself be deceived by ruses.
5. In any form of attack it is essential to assail your opponent from behind.
6. If your opponent dives on you, do not try to evade his onslaught, but fly to meet it.
7. When over the enemy's lines never forget your own line of retreat.
8. For the Staffel: Attack on principle in groups of four or six. When the fight breaks up into a series of single combats, take care that several do not go for one opponent.

takeoff. As per rote behaviour I was flying a gently weaving path – "flying crookedly" – quarter-rolling to observe my entire stern sector every 4-5 seconds and never once ignoring any part of the sky. Yet, a 109-E managed to creep up on me:

The first notice I got of his attack was a stream of tracer flying past my wing and canopy: I could hardly believe I was being attacked. As I rolled and pulled somewhat to avail me of my incredulity, I spotted the bandit no more than 300 meters away right on my tail and slightly high, and immediately accentuated the manoeuvre into a proper evasive action. What luck that he was a right bad shot and hadn't closed in further before opening fire! I had some steam up already so I went into a rolling scissors and was soon gratified to see the 109 flash past, whereupon I immediately pulled in after him. He made a gentle zoom away and from the way he handled his crate I knew he'd lost sight of me. Interestingly enough he didn't make much of an effort in regaining visual but chose to carry on – a fatal mistake, for one cannot fight that which one cannot see.

Fortunately enough his closure had not been massive and I managed to eke out a couple of cannon hits as he slowed down at the top of his zoom, hitting his rear fuselage from low seven o'clock. I could see him shudder with the hits and then go into a long dive, still not seeing me, which I followed.

Having dived to 1 km he then went into a steep zoom, allowing me to cut across his circle and draw a bead as he topped out – surprisingly enough he seemed to be flying at reduced RPM too. My cannon hammered him again, slicing through him from low astern and instantly setting his plane on fire.

The whole combat took no more than a minute at most and beside the initial bounce I was in control of the entire fight. Had he but disengaged after the first missed pass he would have had no worries – but he lost sight, and thus lost the fight. Still, the fact that he managed to surprise me was disconcerting!

The great thing about MMORPGs is that you're always fighting other people rather than predictable artificial intelligence drones. In that respect, every engagement is unique. It also means that in order to excel you have to accurately gauge the bandit's experience and act accordingly. It's not that difficult either – everything the bandit does, or does not do, gives you all the necessary clues as to his level of situational awareness, plane handling capability, energy and ACM proficiency, his mission, confidence and tactical plan. All you need to do is watch him closely. Then of course, you might be set up by a cunning foe who lets you believe that he's less experienced while waiting for you to commit the one error he's expecting. Everything isn't always what they seem to be, and there are more than one way of surprising the unwary. That said, human beings are usually infinitely more predictable than AI drones: their behaviour is predicated by stimuli which, once you figure out how to apply it, becomes utterly predictable. Taken as a whole the enemy exhibits all the attributes of the flock, of the lone stalker, of the vengeful, of the hurried, of the blinkered. Learning to discern the typologies and roughly how and where they appear is very much part of the engagement equation.

Psychology is arguably one of THE biggest factors in air-to-air combat, something that I discuss at length in Part IV. For now, keep in mind that there are no absolutes: your enemy may act rationally or irrationally; he may be experienced or a rookie; he may be hell-bent on completing an important mission or just out for a bit of fun; he may not have seen you or is disregarding you on purpose.

Accurately assessing the enemy's mental state, identifying with him as it were, reading his mind from the way he handles his plane and from the decisions he makes, is in my opinion one of the most gratifying elements of online combat. The moments when you "read" an enemy often contain a measure of glee, which isn't particularly pretty but nonetheless funny – recognising the enemy's fear or surprise or lack of skill, and capitalising on it, is unavoidably material to combat. If you're big-hearted, maybe you feel compassion or pangs of sadness when you gun someone down. I don't!

In any event, the people are what makes or breaks the game. The sense of belonging to a community, of supporting a positive community spirit, is perhaps more important than any individual exploit in the game. To illustrate, you might be a Sierra Hotel gunfighter, but if you lack humility or engage in disruptive behaviour online, you will probably feel very lonely in the game. While building a reputation, self-esteem, or seeking respect from your peers may or may not be important to you, the same rules that govern acceptable behaviour in the real world also applies in the game. In other words, insufferable egomaniacs, people using offensive language, spammers[1] and other high-strung dramatis personae are seldom tolerated very long.

In the air, you will witness great exploits, fantastic scenes – and unfathomable displays of ineptitude. I am ever in awe of pilots who seemingly without effort clear out a whole gaggle of enemy single-handedly, much as I'm disturbed by the near total lack of professionalism in some cases, such as when a whole group of fighters meticulously organised and deployed, dive in and become embroiled, stationary, and defeated in detail. In short, there are all kinds of players out there, and you never know how good they are until you take them on and meet them eye to eye. What a way of socializing eh?

[1] Players who abuse the text interface

CHAPTER 3: FEAR OF DEATH

"We only thought about going out and getting drunk when we got back from the mission - we never considered the fact that we might NOT be coming back."
S/Sgt Leonard J. Hurley, 305th Bomb Group (H)

"I was always scared--that was what made me move quick."
Major Robert S. Johnson, 61st FS, 56th FG

Death, in the online game, means nothing. And then again, it does. The simple fact that you can respawn immediately and go on out to repeat the same mistake over and over again without permanent penalty makes for a warped world when it comes to pilot behaviour. On the other hand, no one likes "dying" for the inconvenience, disruption and confidence-degrading of it. That said, some guys just don't care and even regard dying as an expedient to avoid a troublesome and tedious flight home. Others will go to any length to avoid virtual death – by any means necessary including "hugging the ack"[1] and even "yanking the cable"[2].

Fundamentally, however, players usually rate success higher than failure – if they didn't, what reason would they have to play? In the online game, fear of death is thus a highly individual choice – whether to immerse yourself in the overall experience or to simply regard the game as a simple diversion, a sort of physical "twitch" exercise.

Death means failure. Death does have a meaning in the game insofar as it gives the enemy a shot in the arm, insofar as it increases the odds against your remaining team mates, insofar as it depletes the available aircraft for your side (when attrition is modelled, as in WWIIOL), insofar as it forces you to spend time trucking back to the fight etcetera. But the most important thing about death, which escapes most pilots, is that it reveals your errors. Death is a learning experience which few takes to heart. Instead, the thing most pilot cadavers feel is anger. You get mad – or rather, hopping, screaming, wall-punching, monitor-crushing mad – when you're shot down, a rather disagreeable sensation if it happens with any regularity. Yet this anger is a good thing if you stop to think about it. You shouldn't get mad at the guy who shot you down, or curse the tree that collided with you (heh), but thank them for the lesson and scold yourself for being such a dummy that let it happen in the first place. And then work to avoid death by that cause in the future.

Flying as if your life depended on it gives the game a whole new meaning. Whereas you would normally hop into any fight, you now pick your fights carefully and learn to appreciate the dread and fear experienced by those who literally laid

---

[1] circling the automated anti-aircraft guns for protection

[2] terminating internet access to cheat the opponent on his kill

their lives on the line. Your situational awareness improves, you learn to check your six more often, you learn to maintain energy and positional advantage, and you may even learn how to fight as a team.

Death is the result of sloppiness and abandon. For the purpose of this book, we will assume that the pilot is imbued with proper fear of death, or if you will, an abhorrence of failure. For without fear of death the path to enlightenment is long and painful indeed.

The pilot must not, however, fear death so much that he becomes paralyzed or that he fails to capitalize on advantageous situations. There can be no gain, no improvement, without risk. Fear of death must not be equated with a defensive posture, rather, the pilot should respect death and learn to recognize the envelope of death – in other words, learn to realise his limitations and to know when said limitations are about to be exceeded. Thus, a fighter pilot learns to know himself, and to know his enemy.

It's easy to see who fears death and who does not. The pilot who fears death adheres to fundamental principles of air to air combat: he secures an altitude advantage; he keeps his energy state high; he scans effectively; he flies with a wingman at the very least; he engages with selection; he disengages before the situations turns against him; he uses surprise if at all possible – in short, he hoards as many advantages as possible and never lets his guard down. When this fellow makes a mistake, it's generally an infinitesimally small one: an inopportune stall, a break too gently executed or one false turn too many. If he's worth his salt, such an error is easily recognised and repaired.

The pilot who doesn't fear death ignores every rule ever written. He flies low on the predictable route; he seldom scans effectively; he rarely flies in formation; he engages with abandon and against any odds; he never disengages – in short, he forgoes all advantages for the purpose of getting to the fight without delay and of mashing the trigger as much as possible. This pilot does nothing but mistakes from the time he takes off to the moment when he's shot down.

The learning experience is greatly hastened by taking due note of every sortie and every death in particular. Writing after-action reports, playing the engagement over and over again in your mind's eye in search of the critical error, is a great way to improve. While the terminal error is usually easy to see – overstaying one's welcome, atrocious gunnery, target fixation, hesitation, blackout, collisions etc – the critical error generally precedes the terminal one, sometimes by several minutes, and is usually an error of judgement: you engaged where you should have disengaged, you thought he had less energy than he actually had, you overrated your proficiency or had too little respect for the enemy, and so on. Next time you won't make that mistake again. Next time you'll fly faster, higher, closer, aim better, pull less gravity loads, break off in time, bring friends to a duel.

Fear death, and you will be fearsome.

Sometimes it takes a long time to learn, and some lessons are never learnt. Below are some excerpts from my sortie log back in 1997 and 1998, clearly I had some difficulty with a whole catalogue of basic errors!

Sortie 1: Engine of my 109K smoked by gold spit. Landed.
Sortie 2: Engine smoked again off F13 by vila in pony, extended but crashed against unexpectedly appearing mountain while attempting to ditch under fire.
Sortie 3: In pony. Died low running from 8 golds to F17.
Sortie 4: Pissed. Up Spit5 from F17 in face of heavy gold opposition. Shot two Spit9's and landed.
Sortie 5: Up 109K. Found webs in a FW off F13. Fought for several minutes before gaining the upper hand. I blew his vator and rtb'd w/o 30mm slugs. Webs landed successfully.
WarBirds, September 15 1997

Sortie 1: -ok---(SpitIX), -bex-- (FW), tos--u (SpitIX). All kills delivered while sweeping wink's six. I crash into someone or other and have to swim all the way back to F21.
Sortie 2: -bex-- (FW), -bex-- (FW), -omud- (Bf110). Crashed into E/A.
Sortie 3: Jettisoned ordnance (1000 lb bombs) at low alt to engage enemy 110 south of 21. Killed by own bomb blast. Subsequently bombed and strafed on runway. P/O'd.
Sortie 4: smokey (P39). Pieces of various 110's and B17's. Landed.
Sortie 5: Two enemy 110's accounted for. Die in vicious flat spin.
Sortie 6: Killed by ackweenies at F4.
Sortie 7: In P39, eliminated by -sach- in F6F over F4. No kills.
Sorties 8 and 9 in SpitIX: -bex-- (FW) and various others, log incomplete. Both sorties landed with empty magazines.
Sorties 10 to umpteenth: heavy to F4, log incomplete, but no deaths.
WarBirds, March 12 1998

Sortie 1: outflown by buzzer despite initial altitude advantage. Death by dweebery.
Sortie 2: death by dweebery, general kind.
Sortie 3: engage 110, scoop out low and get too close, touch wing to wing. Death by dweebery.
Sortie 4: flaps extended, throttle idle, bounce from 17K to the deck on Hurri, waste E and continue stallfight despite clear disadvantage. -vert- (?) in the Hurri rapidly outturn my hapless kite. Death by dweebery extremis.

Sortie 5: lead turn Spit but rapidly lose advantage, if any, blown away by buzzer again. Dweebery of a particularly bad kind.

WarBirds, March 24 1998

Ouch!

CHAPTER 4: INTERNET AND MECHANICAL PARTICULARS

It is helpful to have an understanding of how the internet works and how it affects gameplay, as well as to understand that your mechanical setup may increase the odds against you. The online game is an exchange of information between clients (the players' computers) with the game host as intermediary. Everything pertaining to your situation – airspeed, attitude, heading and bullets – is sent to the host in internet protocol (IP) packets, normally one packet every 200 millisecond or so. That information, and the information of everyone else in the neighbourhood, is then sent out in a packet to your opponent (and everyone else in the vicinity) who receives it in the next 200 millisecond bracket. Sometimes the delay is longer, meaning that your data may be up to half a second or more old. That half second is a lot in combat.

While the enemy you see turning in front of you appears to be in a certain place, he is actually about half a second further ahead in his turn. That in itself doesn't imply that you should aim your guns at the spot where he actually is (in terms of internet delay), only that what you see is not the same as what the enemy sees. Consider a turning fight where you see the enemy straight across the circle, in your top view. You would think that neither of you has the advantage. In reality, the enemy (depending on his turn rate) may already be pulling in to a firing position, he may already be firing his guns. Next thing you know, his front end registers hits on you, the host confirms it and you tumble down on fire, screaming "how did he do that!" all the way to the ground. The only way to deal with it is to make allowance for this tiny lag in your engagements. Lag effects is particularly vicious when closure is high, as in when you're travelling at high speeds toward an enemy. Internet delay is most conspicuous when you're parked on someone's tail and pump them full of lead: sometimes it takes several full seconds before the catastrophic damage delivered by you is acknowledged by the recipient, causing over-expenditure of ammunition.

Sometimes internet delay (lag) is so bad that the enemy is seen doing the most outlandish manoeuvres, zipping this way and that or zooming to impossible altitudes. Take it as a cue to distance yourself from that particular fellow, and if it happens repeatedly, as a cue to check your own internet connection. The rule is: if just one plane is seen lagging, his connection is not up to snuff; if everyone is lagging in and out, it's your own connection that is at fault.

Other mechanical effects governing combat are those of your computer and your controller setup. A slow-poke computer with inadequate RAM or dopey graphics card will increase lag on your end, to the point of making air combat a totally unsatisfactory slideshow. It's rough, but in order to fly you must have a high-end machine and be diligent about keeping it fit for fight by regular defragmentation of the hard drive and by continuous hardware upgrades. There's no doubt that online air combat is an expensive hobby!

The flight controller (joystick) is also crucial. Investing in a state-of-the-art joystick (approximately $100-$140) is a wise move if you're serious about flying. A sloppy stick with a bad dead-zone, or a stick which requires overmuch strength to move the control surfaces, significantly lowers your chances against a pilot who flies with a precisely calibrated and well-sprung stick. In addition to the stick characteristics, you should also consider the placement of controller hats, buttons and knobs relative to your hand size. If you have girly hands, don't buy a large and heavy stick with many buttons that you can hardly reach.

Additional controllers – throttle and rudder pedals – are highly recommended not merely for the immersive quality but for ease of play. Having throttle and rudders on a single joystick may seem comfortable, but it's a far cry from having separate controllers. A single hand simply cannot manipulate or effectively coordinate all aspects of flight, other than perhaps for simple cruising around. The main drawback of having everything on your single stick is that whenever you twist the handle to work the rudder or use your thumb to spin the throttle you may also compromise control of the other pitch and roll axes, i.e. you lose fine control of the plane and perform involuntarily. It's much the same with using a Point-of-View "coolie" hat: when you thumb it in one direction, the hand wants to push the stick in that direction too. It takes a conscious effort to remain in control of your plane before you get used to flying on feel alone. There are of course ways of practicing control, one of my favourites being that of flying "backwards": hold a half-back view and fly around, preferably quite close to the ground, without augering in. When you can master that, try taking off and landing while flying backwards too. Another big thing about joysticks is the ease of which you can throw your crate around. It doesn't take more than a flick of the wrist to execute what in reality would be a highly taxing manoeuvre involving your whole body and both arms! Thus the REAL grognards build their own joysticks – full length, floor-mounted jobs that is certain to give you a real workout and a much better idea of flight.

Get to know your controllers before you attempt any fancy stuff, and learn to fly without ever taking your eyes from the monitor – the more you have to glance down to type in commands or work knobs and buttons, the less in control you will be and the more you will suffer in combat.

PART II: FLIGHT AND COMBAT BASICS

CHAPTER 5: THE PLATFORM

The purpose of this book is not to go into minute detail about aerodynamics but into the nature of combat. The reader is assumed to have a general understanding of flight; how lift is generated and what the various control surfaces are for. Suffice it to say that the three axes of movement are Roll (along the longitudinal axis), Pitch (nose up or down) and Yaw (slew left or right). Other than that, two important details to keep in mind are the aircraft's thrust-to-weight ratio (TWR) and its wingloading in comparison to the opposition.

A heavy aircraft with relatively weak engine, such as the Dewoitine D.520, has a low TWR, whereas a light aircraft with a powerful engine such as the A6M "Zero" has a high TWR. A high TWR is naturally desirable, as more power means better acceleration, more lift, better climb, better zoom and a better turning capability. Wingloading is derived from weight divided by wing area. The fighter with the lower wingloading is the more agile, but this is often at the expense of having a low top speed, lack of pilot armour and lack of punch. High wingloading means that the fighter cannot turn as well and stalls out sooner than the fighter with a low wingloading – compare for instance the difference between the Hurricane (low TWR, low wingloading) and the Focke-Wulf 190 (high TWR, high wingloading). These traits are important to remember inasmuch they dictate how to best employ a particular fighter, e.g. the Focke-Wulf is not primarily an angles fighter but should be restricted to energy fighting.

The singular purpose of the fighter is to close with the enemy and gun him down with as little risk to himself as possible. Everything else, in the words of Baron von Richthofen, "is rubbish". To that end, the fighter is equipped with forward-firing guns and/or cannon, the object being to bring these into such a position as to score a quick kill.

At one point of the spectrum it seems pretty straightforward: motor in behind the enemy, kill him dead and get out of Dodge. On the other end of the same spectrum, it's a whole science of situational awareness, flight envelopes, energy states, lift vectors, angles and gravity loads. Since the simplistic approach usually gets you killed in a hurry, it's opportune to learn something about the science and psychology of manoeuvring.

It should also be noted that the platform itself means little from a tactical point of view, other than that it may or may not impose certain restrictions on the pilot's freedom of manoeuvre in various situations. What matters are the conscious and subconscious decisions taken by the pilot, and his moral fortitude. To illustrate this, consider the situation where a supposedly superior aircraft is defeated by a supposedly inferior aircraft because the former failed to secure advantages in the form of altitude, airspeed and surprise, whereas the latter did. Given certain circumstances even a doubly inferior aircraft will prevail, if the pilot knows his own and his plane's limitations.

The platform is important to consider, for it dictates your options and decisions. If, for instance, your rear-view is less than stellar, you will have to manoeuvre all the more to keep your six clear, and preferably bring a wingman to fly line abreast with (more on this later). Other governing factors are rate of climb, rate of turn, rate of roll, turn radius, dive performance, maximum AOA (angle of attack), top speed at various altitudes, acceleration, pilot armour, fuel tank locations and their integrity, armament, drag, engine type, stall behaviour, ruggedness of construction etcetera. Each of these factors are important in combat. Remember though that even if your crate is found lacking in ALL of these respects, it is still the PILOT who makes the difference through his awareness of imposed limitations and through his decision-making. Let's look more closely at the most important of these factors:

Cockpit visibility

If it's good, no worries! If it's restricted, you must roll, bank, yaw and turn to various degrees in order to check your stern aspects. Bad cockpit visibility makes for an increased workload for the pilot, and enhances the need for an extra pair of eyeballs. Still, even under the best conditions, ALL planes have their blind spots: below the nose, below wings and below tail. Use this against your enemy: attack him from such quarters as he cannot easily observe. This naturally implies a fair level of knowledge of the enemy aircraft types. Knowing whether the 109 is half-blind in his low left or in his low right astern may be the difference between a successful bounce and a failed one. Most importantly, if your cockpit visibility is restricted, you must not acquiesce with the situation and take it as an excuse for not looking around you or for not taking appropriate steps as to lessen its impact. Bad cockpit visibility is not a valid excuse for being shot down!

Good: Yak-3, A6M, Bf-110, any bubble-canopy equipped fighter such as Typhoon, P51-D etc.

Bad: Dewoitine D.520, P47 razorbacks, F4F.

Rate of climb

A decent rate of climb advantage (better TWR) over the enemy allows you to use the vertical dimension to disengage from a potentially nasty situation and to close with a slower climbing enemy. Naturally it also reduces the time to combat altitude and it generally saves fuel too. If your climb rate is superior, no worries! If it's deplorable, you must be extremely wary of engaging in vertical fights and wary of climbing anywhere near a combat area. A fighter with low rate of climb absolutely must

engage with an altitude advantage, and thus make all it's climbing to altitude from rear fields and in areas where the enemy is unlikely to be encountered, unless it has a substantial top speed advantage which can be used to increase the horizontal separation. A poor rate of climb is a definite handicap, though it is not a crippling one: you just have to make amends.
Good: Bf-109, Spitfire, Yak, P-38
Bad: P40, P47, F4F

Rate and radius of turn
A good turn rate is a highly prized asset, second only to having a superior roll rate. Since many fights tend to degenerate into turning contests, having the fighter with the fastest turn rate and smallest turn radius is clearly desirable because it allows you to gain an advantageous position and enforce your will over the enemy. Turn performance is a complex issue involving every aerodynamic aspect of the design from type of airfoil and wing shape, horsepower output at various altitudes, max AOA etcetera. Whether your plane is a good turner or not also depends on which enemy you encounter, and under what circumstances – so, knowing the performance of your crate and the enemy's, at specific altitudes and airspeeds, is critical – e.g. the supposedly nimble Zero will hardly turn at all over 400 km/h. Do note though, that there are more ways to defeat an enemy than by out turning him. Angle of Attack is most important for sustained turning: AOA is measured as the difference in degrees between your actual (vertical) flight path and the plane's longitudinal axis. For combat purposes, this angle is created when you haul back on the stick from level flight, or when you try to fly level at very low speed. If you exceed maximum AOA for your airfoil (wing shape) it will no longer create lift and you will stall out.
Good: Bf-109, Spitfire, Yak, A6M, Curtiss Hawk, I-16
Bad: FW-190, P47, Bloch 151/152

Rate of roll
Most important of all in a fighter, is its ability to change heading and attitude rapidly. To do so involves the use of elevators and ailerons, both of which should give immediate and precise feedback without application of overmuch force. Having a superior roll rate means that you can change your heading and/or plane of manoeuvre faster than the enemy, and thus outmanoeuvre him (because the angle of bank [sideways tilt] is important for determining heading and rate of turn, the basic components of manoeuvre). This is especially important when you're in a slower-turning aircraft – as long as you hold the roll rate trump, the better turning but slower rolling aircraft must play by your rules. E.g. if you're being out-turned to the left you can quickly switch to a right-hand turn through rolling and pulling on the

elevators, thus to meet then enemy nose to nose and then to put distance between yourself and the opponent by simply running clean away. Roll rate is not constant – it differs with airspeed – and one crate that may outroll an opponent in one situation may be miles behind at a different airspeed. Typically, your roll rate is slower at very low and very high speeds. If your roll rate is inferior to that of the enemy's you are heavily disadvantaged and must at all cost avoid contests where roll rate is central, such as in a flat scissors fight. That is not to say that a slower rolling aircraft must always be defeated, to the contrary, for there are ways of getting around the problem. We'll talk more about that later in this chapter.
Good: P47, P38, P39, P40, FW-190, Bf-109, Curtiss Hawk
Bad: Spitfire, A6M, Bf-110

Dive performance
The ability to pounce unseen on an enemy from high altitude, and the ability to dive out of trouble, is almost as important as a good climb rate or a superior roll rate. Most fighters are well endowed in this regard although some aircraft "locks up" easily from compressibility effects or must be wrestled down with excessive use of trim tabs and elevator pressure, or suffer stiff ailerons in high speed dives. Significant for a plane that dives well is the ease of going into and maintaining a straight dive; that the fighter retains aileron and elevator control at all but the highest speeds (850+ km/h). It is therefore crucial to know how your fighter behaves at terminal velocity, at what altitude you need to ease back to avoid ploughing into the ground, and at what speed your opponent's performance degrades. Mass, wing design and general construction are the key features here. A heavy fighter of sturdy construction and with thick wings dives a lot better than a light, flimsy aircraft with thin or tapered wings.
Good: P47, F4F, F6F, Curtiss Hawk, P40, FW-190.
Bad: Spitfire, Bf-109, A6M, Ki-43, P38

Top speed
Speed is life. Having a substantial speed advantage over the enemy allows you to engage and disengage at will. High speed allows you to minimise your stay in enemy guns envelopes, to convert speed to altitude when necessary and to close unseen on the enemy. Consider the Messerschmitt 262 jet fighter, the epitome of speed superiority, or the significant speed differences between a Typhoon IIB and a Ju-87 Stuka. High speed is not always a blessing however, as it prevents you from making radical manoeuvres and makes you more than a little predictable once the enemy has his eyes on you. Acceleration is another factor that can be decisive in a fight: you may have the fastest job in the sky but if the slower enemy has better acceleration from

slow to mid-range speeds then he'll be all over you the moment you engage in an energy-burning turning contest. Even so, all is not lost if your acceleration is in the pits. Simply make sure to keep a few thousand feet of air below your belly and use it in lieu of raw horsepower, should the need arise.

Trimming your aircraft

In WWIIOL as in most other air combat simulations you have the option to trim your aircraft, for level flight, for climbing and diving, using the elevator trim tabs. Similarly you can trim your ailerons and rudder for minute corrections in the yaw and roll plane. This capability is something which is lost on the majority of beginner pilots but which is *essential* for any kind of controlled flight, both in cruise and in combat. Your aircraft exhibits specific pitch, roll and yaw tendencies at various airspeeds, tendencies which you control by manipulating the trim tabs mentioned above. These trim tabs are small control surfaces on the main control surfaces which balance the same in order to attain an unloaded joystick state (i.e. controlled flight without stick input). For instance, for take off and landing, it's auspicious to trim the aircraft tail heavy (elevator trim up) so as to avoid having to pull back unnecessarily on the stick. If your crate likes to roll to the right due to excessive engine torque you will have to counteract this with adding port aileron trim. And if your kite likes to yaw left due to damage, add starboard rudder trim to offset the increased drag on your port wing.

The greatest utility however is found in the elevator trim tab. The objective here is to trim your aircraft to an unloaded state at the speed you're currently at. Or in other words, if you're flying at 500 km/h but are trimmed for 300 km/h, you will have to push the stick forward to maintain level flight. Instead, you should trim nose heavy (elevator trim down) to lessen the need for stick input. Similarly, if you're trimmed for 500 km/h but enter a turning fight and thus slow down to 250-300 km/h, you will need to trim tail heavy in order to get the most elevator authority, thus improving your turn performance, to avoid fighting at a disadvantage. How do you know if you're trimmed correctly? It's easy: just let go of the stick!

It should be obvious now that in order to maximise your performance and efficiency you need to jog that trim wheel at all times, and to be aware of your airspeed and trim state at all times. The pilot who doesn't trim his aircraft according to the speed he's at is involuntarily saddling himself with a great disadvantage and is usually an easy mark for the discerning eye.

IT'S THE PILOT, NOT THE PLANE

The guy who wins is the guy who makes the fewer gross mistakes.

Lieutenant Jim "Huck" Harris, USN

For a pilot flying an inferior aircraft, it's convenient and natural to blame his shortcomings in the air on the quality of his plane. While certain aircraft are more challenging to fly, or decidedly inferior in performance, guns, what have you, there are always but *always* ways to work around a disadvantage. Even if you're in a slow, unmanoeuvreable, undergunned and avgas-soaked crate, pitted against an enemy who enjoys every conceivable advantage, there are still ways to prevail. All you have to do is to realise that you cannot succeed in any mission, any time. You will have to choose your mission wisely and carry it out with singular determination (and teamwork), with such tactics as to minimise your disadvantages. You must make amends for your handicaps, or suffer the ignominy of having your enemy expose them for you in a most painful fashion.

In an environment where the enemy holds air superiority, has the faster, the more manoeuvrable, the better armed and the better overall performing aircraft and, logically, the better morale, the opposing force must employ superior tactics and exhibit immaculate caution in the fights they choose to engage in. While the lesser force isn't likely to prevail in the long run, it can still account creditably for itself and remain in being far longer than had it waded mindlessly into combat. History is studded with such examples: France 1940, The Western Desert 1941-42, Italy 1943, France 1944, The Eastern Front etc – in all these examples, one air force was significantly outnumbered by a more powerful and better performing enemy yet managed a creditable performance.

How does one do it? When the mechanical advantages are all on the enemy side, one has to secure all other advantages before joining the fray. That means taking off from rear fields to avoid being clubbed like baby seals at the frontline fields; flying in formation to create local air superiority; climbing to superior altitude so as to gain an initial energy advantage; engaging with team tactics versus the singleton tactics of the enemy; to minimise the appearance over enemy territory to the duration of the critical mission only; to engage with the element of surprise and to disengage well before these advantages degrade.

It is you the pilot who makes the critical choices. There's no reason whatsoever to accept a bad fight unless you wilfully put yourself in that situation. You CAN climb to superior altitude, you CAN make at least one surprise attack and get away with it, you CAN choose where and how to fight. You may be in the slowest, most pitiful aircraft ever designed and still prevail against the fastest, the most numerous and the best aircraft the enemy can bring to bear, as long as you make the correct choices.

Central to this engagement doctrine is to regard the sortie as a mission with a very strict objective and to regard death as the only failure to be avoided at all cost. The sortie cannot be one whose only mission is to "find trouble of some sort and we'll take it from there". Fly with a purpose.

The "pilot over plane" discussion is naturally also a matter of individual ability. How else can one explain the fact that no two pilots handles a similar craft in precisely the same way? Pilot A is extremely successful in his plane: he knows how to ride the edges of the mechanical, situational and psychological envelopes. Pilot B is an intermediate pilot: he manhandles the crate, doesn't watch his enemy closely enough, and his actions are marked by frustration. Pilot A knows what he and his plane can do, pilot B does not. In addition to this, the pilots may have different flight controller setups (pilot A has a great stick, throttle and rudder combo, pilot B struggles with a cheap piece of plastic of dubious quality and must use his free hand to manipulate the keyboard). Until such point as pilot B realises the limitations of his skill and the mechanical impact on his flying (the stick) he will blame his crate and call it a worthless piece of fecal matter.

Online air combat hurts fragile egos. In order to improve and to excel one must learn humility. One must operate with an open mind, with infinite patience and with a willingness to learn. Closing oneself off to what's real, succumbing to impatience or worse, persisting in hubris, will only result in an all the more painful disillusionment and failure to enjoy life.

Remember the three core advantages material for success:
SPEED
ALTITUDE
SURPRISE

When either of these advantages is lost it's high time to disengage in order to retrieve it. It is of course eminently possible to fight without surprise on your side, and sometimes it's necessary too. It's also quite possible to fight without either surprise or speed, since altitude can be traded for the latter, but if you're in a mechanically inferior aircraft and still not quite sure of your capabilities… don't. Still, people being what they are, the great majority will trust their prowess before all else even though it may be sorely lacking… so, back to square one: if your capabilities aren't what you thought they were, fall back on trusting on speed, altitude and surprise.

Permeating the flight simulation community is the inevitable and interminable discussion of who has the best plane, of whom is "über" and whom is, well, up shite creek. Plane types are weighed against plane types, roll rates are bandied against climb rates and turn rates, and never will the discussion come to a rest – because the community is ever in flux with new pilots of little learning and/or experience entering the fray, and because beliefs and myths are stronger than facts. Regardless of game and regardless of where in the timeline of WWII the game exists pilots, particularly those who are saddled with a particular allegiance, will be either lamenting or gloating in their crates depending on which side is holding the trumps. Be that as it may, I submit that it is still the pilot who matters over the plane and those who claim otherwise are simply not sufficiently well endowed in any or all of the core dimensions vital to air combat. Take for instance the match between Spitfires and 109s: normally, the 109 drivers will lament the Spits' turning ability whereas the Spitfire pilots will groan over the 109s potent weaponry and awesome dive and climb rates. Enter the Focke-Wulf 190 A series and the Axis pilots will croon with joy and gloat over their superior speed, dive and armament – until they try to dogfight, as most are wont to do. Still, despite winning the dogfights, the Spitfire jocks will complain and wish they had something faster, something with punch, such as the Typhoon Mk IIB or the better performing Spitfire Mk IX. Give them that and the übermunchkins will be back on the defensive, lamenting their hard luck, until they get another upgrade. And so it continues, without anyone save the precious few realising that speed, climb, dive, roll and what have you, matters little or not at all. Remember, the thing is not what you have to fight with, but HOW you fight with it. Below are a selection of recurring gripes you will hear and/or utter yourself at one point or the other:

"My plane can't climb worth a damn!"
So, don't engage in climbing contests! Do all of your climbing well before entering hostile skies, and disengage when the bad guys threaten to get on top. Be patient, and think "discipline!".

"The enemy keeps winning the turning fights!"
The simple solution is to not turn with them! Engage only in such fights as you feel comfortable in, e.g. close unseen from behind or dive in unseen from above (see chapter 17.2). Set up drag-and-bag traps with a buddy or three (chapter 18.6). Use your other advantages such as roll-rate and top speed, and avoid all else. I know that it's difficult to avoid being suckered into a turning fight, but the decision to turn is yours and yours only. Think "discipline!"

"We are always outnumbered!"
So, fight only on YOUR terms! There is always a rear field to launch from. Instead of spending your online time as a bona fide target drone at the forward fields, always with enemy on top, invest a few minutes in securing an altitude advantage from a rear field. Once on top you are ready to rule. Make single attacks, don't bog down in wrestling individual enemy and disengage the moment you sense your energy advantage is about to be compromised. Or better still, collect some 4-8 pilots and work like a team against the solo or gaggle tactics of the enemy. I know that it's frightfully difficult to the point of impossible to gather the requisite amount of pilots and even harder to make them work as a team, but it really is the only option. It requires real and tangible work, mind-numbing and highly frustrating work, yet if you don't want to make the exertion what the blue blazes can you reasonably expect? Think "discipline!"

"All they do is run!"
By Jove, let them! If you can't catch them, why even try? If the bandit is running away it clearly means that he isn't a threat. It only takes a few seconds to realise that you don't have the performance or the energy reserve to hunt down the enemy. Once you reach that conclusion, abandon chase! At the most, a running bandit is a lure or a ruse. If you can get the bandit to run you've won the engagement, leaving you in possession of the field. On the other hand, the bandit may call it a draw as he is still alive, or he may be planning a come-back on better terms. In any case, if the bandit is running scared, you can get him to turn back for another go by showing him contempt and/or disinterest – check out chapter 17.4. And if he's running home to repair or rearm, well, then you have the opportunity to own his sky so as to kill him all the more effectively next time you happen to meet him. In other words, follow him to his lair and stake him out as he replanes.

"Scumbag nitwits keep crashing into me!"
Since beginner pilots usually fly "pipper-on-enemy" exclusively until they learn the concept of separation, ramming is an occupational hazard. Pitch two beginners against each other and they're highly likely to die in a head-on collision (flying straight at your enemy usually results in collisions you know), whereas two intermediate or accomplished pilots are far less likely to collide – because they fly to avoid the collision rather than fly to collide. It really isn't that hard: simply point beside your enemy in a head-on approach and you will not collide. The novice retorts with "but then he gets a free shot!", to which I say: use the Vertical Luke, i.e. pitch up or down out of his plane of manoeuvre in order to complicate the shot beyond his capability. Up close and personal, if you're feeling the collision coming up, slide

out with a stomp of the rudder or relax stick pressure to slide below and behind your enemy. You must realise that the urge to not relinquish position, even to the point of collision, is nothing but a reflex, a gut instinct, that is utterly false and wholly unproductive. We will discuss this again in later chapters.

What do you do when the enemy is faster and the only thing you have is an advantage in roll rate? Speed is of course the decisive advantage – he who has an overhead of speed can run down the slower enemy and disengage from any fight that threatens to go sour. Thus, the slower fighter needs to secure an energy advantage by storing up on potential speed known as… altitude. This again means that you have to climb, climb, climb and maintain a superior position at all times. However, before long you're highly likely to get into the situation where the enemy is co-E or better thus forcing you to fight on his terms, often while you're in the process of climbing and thus unable to secure that vital advantage, because you're already slow. He who has speed normally has the climb and dive advantage as well, although there are exceptions to the rule. Anyhow, here you are, fighting a faster bandit with nothing but your wits and your roll rate – how do you do it?

Your roll rate is of decisive importance inasmuch it allows you to change direction swiftly and thus create or increase separation which the enemy cannot immediately follow or make up for. In short, every time he commits to an attack, you have the opportunity to roll and pull outside his performance envelope. Thus you evade his every pass, flying outside his "cone of opportunity". Every time he misses, he must set up anew by turning around and that he cannot do without a certain separation (distance) as that would put him unhealthily close to your guns. While he turns around you move out again, towards home, friends and AA guns. If you're feeling somewhat more combative, you can easily set him up for a rolling scissor as he comes boring in on you (see chapter 13.8), and polish him off once and for all. For it is highly likely, given his speed advantage, that he has a poor turn performance and is likely to stall out if he handles his crate with abandon.

If found at a disadvantage while climbing you must immediately take steps to ensure survival. If the bandit hasn't already engaged, level off and build up enough speed to manoeuvre profitably against the upcoming attack. If he's already coming down on you, roll away and down at right angles to his heading. As he corrects his course and bores in, roll and pull low again to get out of his guns solution. From there on it's a case of doing what's best for the situation as it unfolds – more on this in later chapters.

The enemy is faster AND rolls better – all I have is a slight advantage in the turn?
Now you're doubly disadvantaged. The enemy is faster and rolls better, giving him the opportunity to follow your every move and to produce excellent guns solutions in due time. You're flying a crate that can turn well, and that's it, so how do you prevail? Well, the enemy will want to fight in long, fast, straight moves with as little turning as possible, whereas you will want to slow down the fight and turn as much as possible. In order to survive and prevail you must fight on your terms only, and you must avoid a pure rolling contest too. That said, your enemy's performance is likely to suffer markedly if you can bring him down to a slow speed, allowing you to work with rolling scissors despite your slight roll disadvantage – for the rolling scissor isn't as dependent on massive aileron use as it is on judicious use of elevators to manage vertical separation. In the rolling scissor, you use your roll to corkscrew around your enemy, not to roll completely away as in the flat scissor. So then, bring him down to a slow fight by flying close to him and by cutting across the circle he makes in an effort to reverse on you. He will try to increase separation so as to turn completely around on you before closing again, if you can deny him that separation he will eventually slow down to fight on your terms or disengage altogether in frustration. The critical part of the fight is when he manages that important separation and comes at you – you will have to feed him angles on his approach and use the rolling scissor against him, for that is the only thing short of a bare-knuckle knife fight that will save you. Check out chapters 13.8 and 13.9 for more detailed information on how to deal with these situations.

What do you do if you are thrice disadvantaged?
You cannot run. You cannot roll your way out of Dodge. And you cannot outturn your assailant. The only thing you have going for you is a slightly more levelled field in the slow fight, as your disadvantages aren't quite as marked then. Your choices are extremely limited: other than knocking down a couple of stiff ones in the airdrome bar, you must fly well above anything remotely hostile and only engage if you can do so with the prospect of killing your target in one blow or of destroying your enemy's SA[1] and energy in that single attack. You may also want to seriously consider bringing along a fistful of friends so that you can set up drag&bag traps for the unwary. In the co-energy or worse scenario, which is bound to occur since you went up alone to brave the odds, your only alternative is to roll your way home while making yourself as unpredictable a target as possible. If you're really good you may get a few snapshots in such a rolling scissor disengagement fight, but don't bank on it. If you do elect to fly and fight (maybe you don't have a choice!) you should attempt to find the enemy where he's low and slow: at or near his base, on climbout

[1] Situational Awareness, see chapter 10

stretches, while returning with battle damage etc. Jump on him then, hack him down and make yourself scarce right quick! Nothing hones skill as well as fighting in a completely disadvantaged aircraft – try it, and you might just get to like it. If nothing else, your kills, when you get them, will be all the sweeter.

Some of the best (virtual) pilots I've known deliberately flew the crappiest aircraft and more often than not never bothered with securing positional advantage, and still managed to inflict serious blows to the opposing ego. Squire Toad for instance habitually flew his big, clumsy, flimsy, slow and under-gunned Aichi Val divebomber into combat with the heaviest American Iron, Spitfires and the like, kicking their heinies. Minister Worr preferred the comparably slower and under-performing P38-F against much faster, nimbler and better rolling enemy, doing more than well for himself. The kick of gunning down a guy in a thrice advantaged aircraft is exceptional. I've waded into (and successfully out of) combat in a Blenheim Mk I fighter-bomber, a ruddy big and unwieldy aircraft with absolutely pitiful cockpit visibility, a plane that was totally at a disadvantage against the opposition in all aspects but turn radius at near stall speed, which is a dubious advantage at best. While these examples are but lore and braggadocio, they do support the claim that winning has precious little to do with the aircraft and its capabilities but very much more so with the brain behind its controls.

We can't support our troops because the enemy is too numerous and have better aircraft than we do.
If I had had a dime for every time I've heard this I'd be a rich man today. It is inevitably so (in a game) that one side will outnumber the other, one side will have a better performing inventory, one side will have the better pilots and one side will enjoy all the benefits of superior morale. What's worse, depending on your allegiance of course, is that these factors are self-reinforcing: good inventory = more pilots = more success = better morale = more flight time = better pilots und so weiter. Griping and yammering isn't going to change that. If you're left holding the wrong end of the stick you had better do something about it, right quick. The first thing to realise in this situation is that you cannot expect to succeed in any venture at any time under any circumstance. Strike that thought from your mind. You can only succeed in such situations as your proficiency and circumstances allow, and with the above-mentioned set of disadvantages ranged against you those situations are few indeed. Tough luck, but there it is. For more advice on this topic, see chapter 19.8.

Back in 1943 Air Vice Marshal Tedder laid down the law on how fighters should be employed, in order of priority and in order of discrete tasks to be accomplished before the next can be contemplated:

a. Fighter sweeps to clear the enemy out of the sky.
b. Escort for light and medium bombers.
c. Interception of enemy aircraft.
d. As a fighter bomber to provide CAS for ground forces.

This doctrine makes all sense. Before you can afford the luxury of CAS[1] you absolutely must win air superiority. To do so, you must find and defeat the enemy in the air and on the ground, destroying him with fire and bombs, with interdiction and denial of production. Such enemy aircraft that venture to do the same to us must be intercepted and destroyed, and only once these tasks have been satisfactorily completed can CAS become an issue. So, if you're outnumbered, outperformed and outmoraled: fly higher, fly faster, engage with distinction and disengage at the first whiff of advantages lost. Never ever fly alone, treat your team with as much care as you want them to show you and learn to dominate locally before shooting for total air superiority. What does all this have to do with "pilot over plane" you ask? Well, it is not uncommon for sim pilots to make too steep demands on their rides, and on themselves, that they cannot deliver. Realising one's limitations and adapting to them is probably the single most important factor in surviving a sortie.

[1] Close Air Support, i.e. air-to-ground attack in support of friendly troops

CHAPTER 6: ENERGY

Speed is life.

Unknown pilot

The concept of energy is absolutely central to air combat. Without an understanding of energy or how to deal with disparities of energy under various circumstances, no fighter pilot can hope to survive. When talking about Energy, or simply "E", it's generally assumed that maintaining a high energy state is the way to go, and that if you have it, you can't go wrong. Wrong. Maintaining a high energy state (flying higher and/or faster than the opposition) is not the be all and end all, although having the big E is certainly preferable to not having it. What matters is the comparison between your energy state and that of the enemy's, and how these energy states are used. Thus, energy is always relative – and, having a high energy state is not always a blessing, as we shall see. That said, the incontrovertible truth is nevertheless that he who holds the energy advantage dictates the fight for he is free to engage and disengage at will.

Your energy state is the sum of your airspeed and your altitude (to be totally true energy is the sum of airspeed and position relative to the enemy) as created by your powerplant and conscious choices. Either of these two factors can be traded for the other: airspeed can be traded for altitude, which can be traded back to airspeed. When you ram that throttle forward you're increasing your energy. When you convert some of that speed to altitude by climbing you're essentially "putting money in the bank" to be used at a later time. The wise pilot always maintains a positive bank account! He will need it to execute surprise attacks, to manoeuvre against sudden threats, to close distances rapidly and to disengage safely.

Note that a fast fighter at a lower altitude may have the same energy state as a slow fighter at a higher altitude, although the altitude difference cannot be much more than 2000-4000 feet and the high fighter needs to go pretty damn slow for them to be energy-neutral. What's more, (relative) energy is dependent on position and vectors: the energy total of two fighters approaching head on is vastly different from that of two fighters pointed away from each other. Consider a meeting engagement between two co-E, similarly capable, fighters: if one of them opts to blow right through in the merge his opponent must spend a certain amount of energy and time to convert to his tail, and by then the disengaging fighter is likely out of guns range. Thus, your position and vector relative to the enemy is of great energy value and as such needs to be assessed at every juncture.

The energy equation of speed+altitude+vector is what allows a patently slower fighter to dominate a technically speaking faster fighter. On paper the slowpoke plane shouldn't have a chance, right? Well, with a bit of altitude to be cashed in for speed and distance (from an advantageous position I should add) almost any plane can be brought to bear against a supposedly vastly superior fighter. What's more, that faster fighter isn't always fast: he may be cruising at reduced power settings, he may be handicapped by battle damage, he may be turning, climbing, circling, landing or in a number of other ways remaining essentially stationary. In such a case his top speed matters little. As in real estate, position is all that matters.

As stated above, if you have a significant energy advantage you lay the rules of the engagement. If the situation is or becomes unfavourable, you may want to disengage or seek a better position from which to engage. Until the advantage is no longer significant, the low-energy fighter must seek to meet the high-energy fighter at the best of given terms and to work towards levelling the respective energy states. Typically, the high-energy fighter will engage with the altitude advantage and stick to moves that burns as little E as possible – in order to keep up his attacks with minimal risk to himself for as long as possible. Conversely, he wants the enemy to stay disadvantaged: he will want him to burn energy by forcing him to engage in heavy defensive manoeuvring. Another way of burning your opponent's energy is to make him lose situational awareness – he will have to burn more energy in looking around and keeping his six clear. Thus we can speak also of situational energy and morale-related energy.

Burning hard-earned energy is naturally extremely bad. Energy is destroyed when airspeed decreases without a consequent increase in altitude; by increasing drag (through pulling excessive gravity loads in turning, by using flaps or excessive use of rudders); by deliberately chopping throttle; by skids and through dumping of altitude without consequent increase of airspeed. Look for these telltale signs in combat, and avoid committing them yourself. In other words, never slow down in combat. You will be doing yourself an immense disservice if you pin your survival on a bag of tricks such as use of combat flaps, snap rolls and tail slides – before all else, learn the basics!

The successful pilot hoards energy and hangs on to it for dear life, never spending more than is necessary to keep himself safe or to score a kill. You have to be a miser with energy and always strive to keep energy-bleeding manoeuvres to a minimum, lest you be found flapping around on the edge of a stall without smash to avoid a sudden onslaught. The true mark of an energy-conscious pilot is that he gains or at the very least retains energy where others spend theirs, making his flight seem effortless and ever unpredictable. As long as you have energy, you have options.

An overhead of energy – lots of speed and lots of altitude – is a cushion against sloppiness. If you have a significant energy advantage, you can afford making a bad attack or two, though probably not more. As long as you keep this advantage all the enemy can do is to defend himself and make himself a difficult target.

You must be ever mindful of your energy state and keep a "balance book" of speed, altitude, vectors and potential moves in your head. Furthermore, while you may have a superior energy state against the one bandit you might soon be confronted by more whose energy states are vastly superior to your own. We call these bandits "wildcards" – unknown entities – which may at any time enter the equation and ruin your fun. To this end, it helps thinking of energy in different magnitudes: you have the "big picture" which spans the whole engagement area beyond visual range, you have the "first contact" energy states in the engagement and you have the "close in" energy states when you ride the edge of your own and your opponent's ability. It is important to consider each of these in turn, and conjointly when discussing a fight in particular.

As a rank beginner, this all sounds like mumbo-jumbo, right? Well, contrary to popular belief, a good fighter pilot is not the one who throws his crate around willy-nilly although it might *look* fancy, but the one who makes small, precise and economical moves, spending only so much energy as is required to gain an advantageous position. Ideally in a sortie, you should never have to dig deep in your energy account through wild maneuvers and expensive turning, but operate at a happy, balanced state at all times. If you are always found scraping the bottom of the energy barrel, using WEP[1] at treetop altitude and burbling on the verge of a stall, do realise that you're in dire need of rethinking your whole approach to aerial combat.

That is not to say that all-out turnfighting is bad per se for surely there will be times when no other option is available. However, to engage exclusively in energy-poor manoeuvring while casting virtually all other options aside, is about the least productive strategy for learning a beginner can choose. In order to master such a fight, which is at the very extreme of the fighting spectrum, one must first master the whole.

THE "BIG ENERGY" PICTURE

When you're cruising around looking for a fight, you will want to do that at an energy state which allows you to dictate the engagement from the word Go, rather being forced to deal with the opposite at every turn. In other words, you should cruise several thousand feet over what is normal for the general area, and at a good clip too. The higher you go, the more you can relax and cut down on boost and RPMs to save fuel and extend your stay in enemy territory, while still being relatively certain of holding the energy trump. As long as you're on top you don't necessarily have to motor along at top speed as this wears out your engine and burns gasoline for precious little gain: it's a small matter to power up and spend a few thousand feet to get up to combat speed without letting go of a sizeable energy advantage. You can

[1] War Emergency Power

safely assume that anything that moves lower and slower (or both) than yourself can be adequately handled as long as you manoeuvre into a favourable position and take advantage of the situation. If you eschew this – securing a superior energy state – you will always fight at a disadvantage. Voluntarily choosing a bad starting situation defies logic.

Whenever you engage in a fight and deplete your energy, you must consider the "big energy picture" for that particular area. Even though you may have retained a good deal of energy it may be drastically lower than that of your typical wildcard in the area – if so, you must put some distance between yourself and likely entry vectors so as to safely regain energy in a less trafficked area. In this regard it should also be noted that there's a measure of safety to be found in lurking in the weeds if the bulk of wildcards can be assumed to be found at much higher altitudes. It's a crapshoot though, for it only takes one vigilant soul, or one lucky sod, to get on your six and draw the attention of a whole host of bad guys. And that can really ruin your day. In other words, you may be stranded with a relatively low total energy in an area where the bulk of wildcards have a relatively higher energy. You certainly don't want that – exit this unhealthy area without undue delay and regain a reasonable energy state.

"FIRST CONTACT" ENERGY

Assuming that you encounter a bogey at roughly the same altitude and in your forward quarter while you're cruising around at a good clip, the relative energy state is largely unknown until you engage in combat. Still, with experience, you should be able to deduce a lot, well before you actually meet. As soon as you lay eyes on the bogey you should be computing and comparing: your own speed; difference in altitude; heading and angle off tail; rate of closure; type of bandit and whether he's seen you or not. If you're approaching head on the likelihood of him not seeing you is small (although it happens!) and the only thing that really gives you a clue is rate of closure. Hence, from an intelligence-gathering point of view, a head-on approach is about the worst possible case.

If your intention is to engage rather than to disengage, it is far better to introduce a measure of lateral (horizontal) separation by pointing your nose well to either side of the bandit. You already know that you've seen him, and the lateral displacement is your first offensive manoeuvre – question is, has he seen you, and what will his response be? And more importantly, what is his energy state?

Another way of dealing with the same situation is to add a vertical component as well (i.e. dive or climb slightly). While this robs you of some altitude or airspeed, it doesn't materially degrade your energy, to the contrary, and will give you answers to the same questions above.

The beginner pilot usually doesn't consider any other move than the one which brings him directly towards the enemy – i.e. he will always point his crate straight at the enemy, and when the enemy zips by (provided that he doesn't collide), he'll haul around as hard as he can and point at him again. Needless to say, that approach is more than ineffective, it is also totally predictable. This type of fighting doesn't include any intelligence-gathering or tactical finesse whatsoever, it's merely a question of pulling G's and flapping about in the vain hope of getting guns on enemy.

Now, at first contact, the objective is to make an instant assessment of relative energy states and to arrive at a game plan for the ensuing engagement – other than simply hauling in on the bandit. You need to be relatively sure of whether you have the advantage or not, and this process of collating information is concurrent to manoeuvring into a favourable position. Again, instead of trucking right into the fray, you secure an element of lateral displacement by pointing well to the side of the enemy. His reactions to this move tells you a lot about his general awareness and intentions, as well as his energy state. Assuming that you're roughly co-energy and that the bandit does the predictable thing of turning towards you, you have started what will turn out to be a lead-turn manoeuvre. At this point it seems like the enemy has the better position since he has his guns pointed at you, however, that is merely to appearance. At the time when the bandit thinks himself set up for a high angle off snapshot, you turn into him and manoeuvre around on his six with a vertical reversal (i.e. you make an oblique 180 degree turn: following the turn into the bandit, pitch up as for a loop, then continue the roll to upright as you come over the top and start heading down), taking advantage of the gravity assist to slice in across his predictable turning manoeuvre. Note that this is merely one of many possible gambits you can choose from.

Once the opening move has been made you should be fairly aware of the energies involved, and the proficiency of your opponent. You're now free to work your way into a favourable guns position. We'll get back to the actual manoeuvring in later chapters.

These "first contact" energies can be more or less easy to determine, the level head-on approach being the most difficult of them. It's a whole lot easier to determine relative energy if the enemy is moving across your front, and especially so if he's trucking along at a lower altitude. What you need to keep in mind is that the energy you have at the outset of the fight is likely to be it: once you start turning, you'll be forced to burn energy. It is also here, at the beginning of the fight, that the novice pilot burns the most energy by going tongs and hammers after the bandit without regard to whether it's wise to do so or not.

I mentioned position and vectors earlier as one important aspect of the energy game. Consider again the level head-on pass, but this time from a defensive point of view. Now the objective is to avoid the fight, so instead of taking out a lateral displacement you keep trucking toward the bandit – though not straight at him. You will want to pass close enough to avoid him getting on your six yet far enough away to avoid a simple head-on shot. If the enemy is to have any chance of engaging you, he will have to make a lead turn so as to place himself in your rear hemisphere at the time you pass by. This however is beyond the capabilities of most pilots, giving you the opportunity for a clear getaway. What you do is to turn the energy of the bandit against him, knowing that he must spend energy to get into a favourable position (on your six). If he times his manoeuvre badly or doesn't manoeuvre at all by the time of the merge, you will be increasing the distance all the while, speeding out of his guns envelope as it were. By the time he hauls his crate around you're not only far distant but have kept your original energy as well, whereas he has lost both position and energy.

"CLOSE IN" ENERGY

When engaged in a turning contest, be it nose-to-tail or nose-to-nose or a vertical fight, popular belief has it that energy is not a factor, that only angles matter. Nothing could be farther from the truth. Energy is ever important. Energy is what makes or breaks a fight, that which tells you whether you can or cannot perform a certain manoeuvre. Keeping track of energy states in the fangs-out hair-afire fight is a real challenge since positions, altitudes, angles and gravity loads keeps changing. Here, you have to think in terms of "level", "uphill" and "downhill". You will have to watch relative attitudes and minute altitude differences very closely and watch especially for signs of the enemy chancing a lag pursuit, whether he deploys flaps, chops his throttle or works extensively with rudder. A careless stall following an optimistic lead pursuit curve, a sloppily executed manoeuvre, a momentarily lost visual contact, is all it takes to eradicate an energy advantage and turn the table right over.

It is eminently possible to increase the energy disparity to your benefit while engaged in the all-out turning contest, much as it is eminently possible and indeed normal to throw energy away. In the "level" and "downhill" instances, a reduced back pressure and less sloppy flying using rudders or flaps will cause you to fly cleaner and with less energy expenditure, and while going "uphill" you can likewise ease back on the stick or opt for a lower climb angle. This of course must be weighed against the very real risk of losing position: normally one "buys" position through energy expenditure. Choosing to stash up on energy (by relaxing back pressure on the stick) at the cost of a few angles at one point may be the thing that allows you to perform the decisive manoeuvre later on in the fight. Therefore, knowing what you can do or cannot do with the energy currently at hand, and what you will want to do with your energy once you recoup it, is the very essence of the all-out fighter duel.

The objective of the close-in dogfight is arguably to get your guns on the enemy with minimal delay. However, you will want to do that with as high a probability for a kill as possible. To this end it helps a lot thinking about energy: you will want to bleed his energy and thus reduce his options to a completely predictable state where he has no option but to wallow in front of your guns. As long as he has energy he can manoeuvre against you: once he's without energy, he's at your complete mercy (naturally provided that you keep yours, and use it to gain a rear-aspect position). This energy-bleeding process is quite rewarding, but it is also accompanied by a measure of risk as the process isn't immediate. If you're primarily an angles fighter, you will have to spend some time with the bandit, get on his six and stay there. This practice is wholly inappropriate for any engagement other than a pure 1 vs. 1 fight, and even then it's debatable whether it's sane to "play with your food". The energy fighter reduces his enemy's energy by other means: by subjecting him to a series of attacks that leaves him no option other than to spend energy through radical guns defence.

Learning how to retain energy and how to make the enemy bleed his is arguably the difference between the experienced and the inexperienced pilot. Little by little, you'll learn to make your moves at as little cost as possible, when it's as most auspicious to do so. Here are some clues:

When at high speed, make small moves.

- Rolls costs less energy than hauling back on the elevators.
- If you need to slow down your forward progress, shed speed in the vertical (i.e. by zooming up) and roll to keep your eyes on the bandit.
- Make your major turns when the energy cost is the smallest – i.e. at or near the top of vertical manoeuvres.
- Break off unproductive passes before you incur an energy penalty.
- In a turn fight, use lag pursuit to get your energy up.

What does "levelling the enemy's energy advantage" really mean?

At some time or the other you have probably found yourself defending against an enemy who holds the energy advantage. He's subjecting you to a series of slashing attacks, and all you can do is to keep turning into him, defeating his passes by presenting him with low-probability shots. Question is, how do you get out of the rut and even out the fight? Most beginners simply stay put under the enemy's guns and even try to climb below him, making subsequent defensive moves all the more difficult and less likely to succeed because airspeed degrades to the point of making any manoeuvre worthwhile. Rather than playing sitting duck below a boom&zooming enemy, you need to cancel out his energy advantage by introducing lateral separation. In other words, as the enemy zooms, you fly away from him, opening up the distance. Next time he attacks, he must first cover the lateral distance before closing to guns distance. This takes time, and forces him to burn energy in a shallow dive from which he cannot hope to regain as much energy as he would had you remained largely static below him. As he misses his next attack, you open up separation again. Before long, his attacks will become less frequent, less aggressive and a lot easier to defend against, indicative of energy having been levelled.

Another way of dealing with the problem is to dive out and force the enemy to come after you. The objective here is to reach terminal velocity, where closure is small or minimal, and then to manoeuvre from a more or less neutral energy state (though at an angles disadvantage). As you level off, at top speed, the enemy should be closing but not as rapidly as if you had remained below him. Now the energy game is more or less even: defeat the upcoming stern attack, and either open up lateral separation as above or go into a rolling scissors fight. In either case, you've successfully levelled the energy equation. Also, chances are that the enemy will be a fraction slow in following your pull-up (or your defensive brake), resulting in a vertical overshoot - i.e. he bottoms out below you. Another thing to consider here is the mechanical performance of your crates at terminal velocity: who dives and accelerates better, who has a roll rate advantage in the dive, which crate handles compressibility better etc. We'll discuss this more in detail later on in the ACM chapters.

Too much energy!

When considering a fight, we generally assume the opposing pilots to manoeuvre close to each other and seek the shortest possible route to each other's six o'clock. However, there are all kinds of situations out there and not all of them favours the fighter with an overhead of energy. Being on the receiving end can actually be an advantage under certain circumstances, particularly the one most dreaded by the average pilot: the high six o'clock attack.

Consider this: you're motoring along and are assailed from high six by an enemy that you've been keeping your eyes on. As he streaks in with a ton of smash, his superior speed is actually more of a hindrance than a blessing: at such a speed, he is extremely limited as to how much gravity loads he can pull and with how many angles he can correct his approach. You on the other hand are flying at a normal cruise speed and can thus introduce more angles into his attack run than he can handle: a simple course correction to 90 degrees off tail is usually all that is needed to defeat his pass. In order to draw lead on you without blacking out, he must slow down and thus burn energy. This is even more true when the enemy assaults you from a flank or from directly ahead – all you need to do is to throw him a few angles, i.e. turn to increase his angle off your tail, and you will see him miss his attack by miles – see figure 1 below. This manoeuvre can be started while the bandit is yet quite some distance away: if he's any good, he'll break off his attack and seek a better position. If he's unskilled or flying with hope as his co-pilot, he'll throw caution to the winds and bore right in regardless, pulling gravity loads and making his shot all the more difficult. The farther out the enemy is, the more gentle you can make your manoeuvre so as to retain your energy. If he's really close, you need to make a hard break turn so as to get quickly out of his manoeuvre envelope. Then and there you have him.

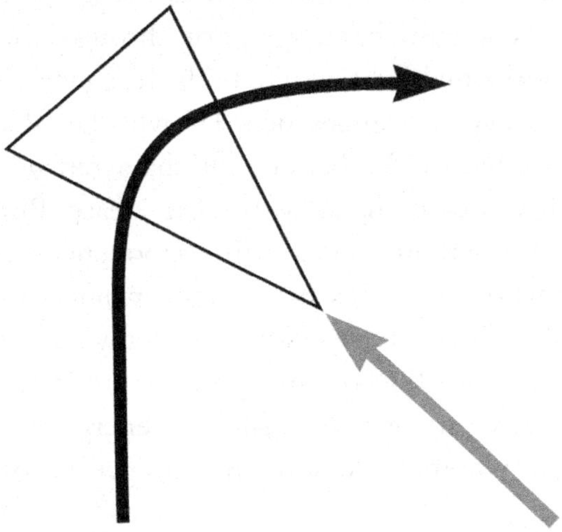

Fig 1. The enemy manoeuvre envelope, top view. The attacking fighter (grey) can only pull so many degrees of deflection without suffering adverse gravity effects, which also impacts his gunnery, and/or suffering loss of lift required to manoeuvre. Ergo, know what your enemy cannot do, and use it against him.

Thus, should you find yourself flying around with lots of enemy well above, you need not fear them overly much, knowing that their overhead of energy will prevent them from making effective passes at you. In that respect it is safer by far to fly a lot (10,000+ feet) lower than the enemy, or just below him (inside his turn radius), than within easy grasp some 1-3,000 feet down. When it's you who have the significant energy advantage, you will have to shed some of it to a manageable level lest your attacks be predictable and easily defeated.

This same overhead of energy is a great handicap when you're coming in with a lot of smash to help a friend who's turning with an enemy at low speeds. While you may close the distance rapidly, you will not be particularly effective unless the enemy is extremely slow or extremely predictable (allowing you to set up an unloaded lead pursuit pass). It is not uncommon to happen into such a situation with more than double or even triple the energy, which naturally restricts you to time-consuming boom&zoom passes. Meanwhile your partner is in deep doo-doo with an enemy who might have a shot opportunity every 5 seconds. Thus, you must shed your energy to a manageable level, or to a factor of 1,5:1 over the bandit. If you burn too much energy you may likewise be unable to saddle up quickly, exposing your friend to enemy fire and possibly even exposing yourself.

Energy in the dogfight

When the chips are down and you're engaged mano a mano with the enemy in a turning contest, energy is almost tangible. It helps greatly to think of yourself and the enemy in terms of angles-rich/angles-poor, energy-rich/energy-poor through the various stages of the fight. In a single-circle horizontal fight, you're both struggling to haul your crates around and to draw lead, to become angles rich. He who has the more powerful engine, or the lower wingloading, or the better stall characteristics (can pull more AOA), he who is energy-rich will earn angles with relative ease. But what if the crates are more or less evenly matched and what if the opponents are both angles-neutral and energy-neutral? He who relaxes his turn, either willingly to exit the circle or through an inopportune stall, will eventually allow the enemy a clear shot. Still, simply staying in a flat nose-to-tail chase that doesn't produce sufficient angles quickly, or at all, or in one which you're clearly losing, is silly. Co-energy, co-altitude and locked into a repetitive behaviour, you must do something to get out of the rut.

The reflexive behaviour of most pilots is to honk back on the stick for all they're worth, pulling as many G's as they can, turning as sharply as they can. This leads to a quick loss of energy and eventually to a state which leaves the pilot with no options whatsoever: he'll be low and slow on the deck, unable to cash in more altitude for speed. It is not uncommon to see two fighters turning flat circle after flat circle without either gaining the upper hand, for minutes on end. The obvious answer to this dilemma is to keep the fight dynamic. Sticking to a nose-to-tail chase is a static situation despite the apparent motion: you need to introduce the dynamic and capitalize on the reflexive behaviour of your enemy.

Your options, when co-energy, to get out of the rut, is to reverse the circle with all its risks and benefits as in figure 2 and 3, or to momentarily go to lag pursuit in order to stash up a smidgeon of energy, with all its risks and benefits, as in figure 4. In Figure 2 below you give the enemy a small but potentially dangerous opportunity shortly after the reversal as you bring the fight nose-to-nose instead of nose-to-tail.

Remember that it's the fellow leaving the static circle that relinquishes position, i.e. loses angles and becomes the defender.

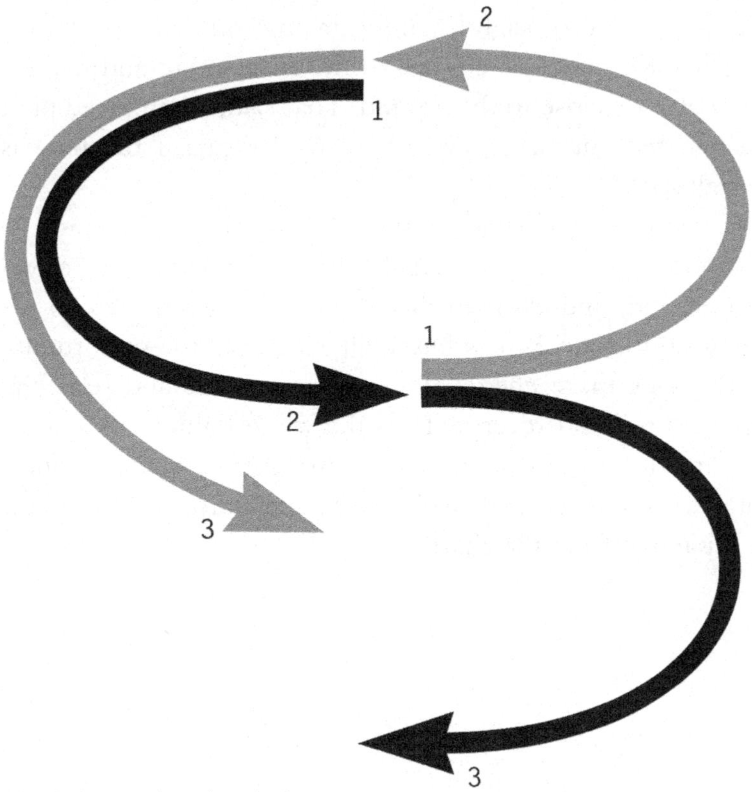

Fig 2. Reversing the circle.

At time 1 the fighters are essentially neutral, locked in a flat, static, nose-to-tail circle, canopy to canopy. The black fighter decides to leave the circle, either because he feels himself slowly losing position or because he simply doesn't like the unproductive situation. At time 2 he reverses the circle by rolling 180 degrees out, briefly blanking out the enemy behind and below his tail. This isn't so much of a problem as one might think – it's fairly clear where the enemy will go, particularly if they're on the deck already – the grey fighter must by logic and reflex continue to haul towards the black fighter.

Note that it's eminently possible to roll either way to get out of the circle. For some reason, probably due to lack of experience and a desire for security, most pilots chooses the rather time-consuming process of reversing upright, nose-over-horizon, whereas the inverted reversal is quicker because of the gravity assist. I.e. in a counter-clockwise circle the pilot rolls 180 degrees right to produce a clockwise circle towards the enemy. The inverted reversal has the pilot rolling "inwards" into the circle until the canopy points fully down and then fully away from the bandit. I.e. in the counter-clockwise circle, the pilot rolls 180 degrees to left, canopy to ground, to produce a clockwise circle. This is more efficient insofar as you need not fight inertia and gravity to reverse the controls (push the stick over) but simply exaggerate the already established bank. However, it is also slightly more taxing on the Situational Awareness and an over-eager pilot may lose control in the manoeuvre and make a mess of himself, especially if already close to the ground. That said, a hamfisted pilot is equally liable to stall out in the "normal" reversal, particularly so if the fight is already conducted at near-stall speed.

At time 3 the black fighter has regained visual on the grey fighter and is manoeuvring to defeat his shot: he throws him a small vertical element (up or down depending on proximity to ground) and trusts in the white fighter's reflex to go for broke. In the interval leading up to time 3 the white fighter has had the opportunity to "clean up" a little bit since it's more energy-efficient to continue and relax the already established turn than to roll and reverse. Thus the white fighter has a small energy and a considerable angular advantage at time 3 and gains a guns position earlier than the black fighter who must struggle to bring his crate around for a shot. This is where the wheat is separated from the chaff!

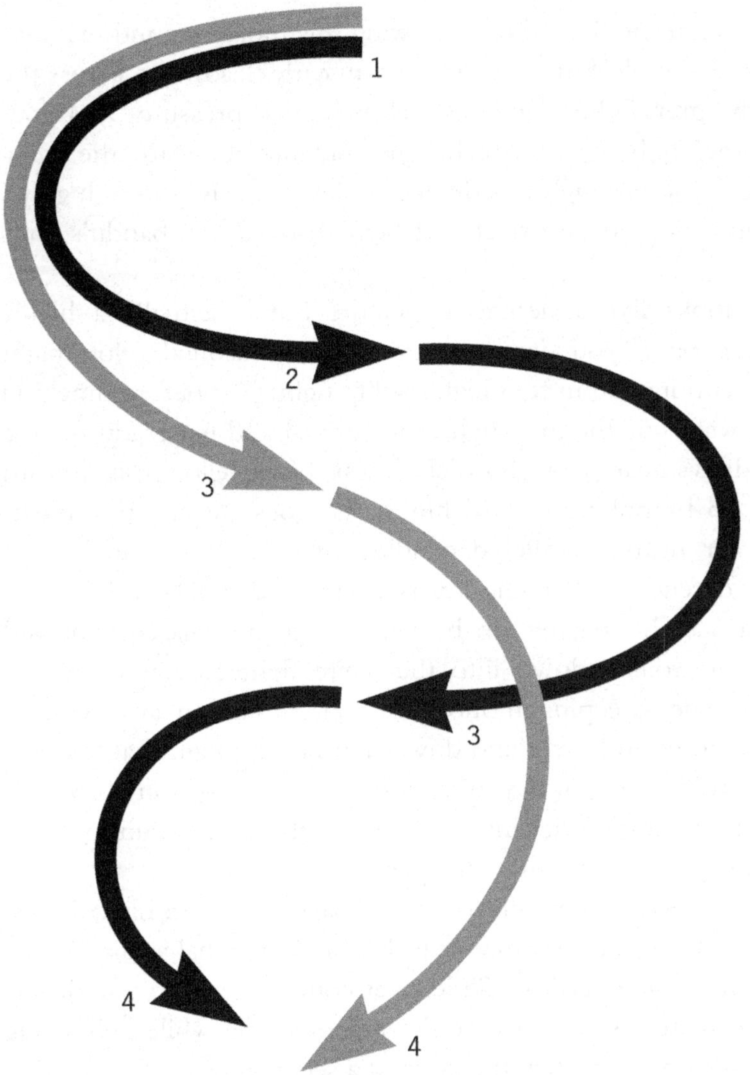

Fig 3. Reversing the circle, continued

Knowing just what kind of a target he's setting himself up as and that the grey fighter, now salivating wildly, will relax his elevator pressure to unload for the shot, the black fighter completes the setup. He knows that he will not have a chance to get a good shot off himself and that he will cripple himself if he tries, by becoming predictable. That, becoming predictable, is what he must avoid at all cost and what the grey fighter is falling for. The way to capitalize on the situation is to treat the guns defence situation at time 3 as an opportunity for a lead turn, a rolling scissor or for a blind reversal (i.e. in the bandit's low cold six), all depending on the angles, energy and separation at time 3.

Distance allowing, i.e. if he has time to swing his nose around without completely killing his energy, the black fighter can start an early (lead) turn across the grey fighter's track. As the grey fighter unloads (releases stick pressure) and goes straight for the shot, the black fighter performs his guns defence and starts the swing onto the grey fighter's back. This is a high-risk defence unless the rate (through grey's guns envelope) is good and the vertical offset sufficient to make the bandit's guns solution tricky enough.

The rolling scissor looks like a defensive posture but is actually a highly aggressive stance, a "come on if you dare" gesture that a seasoned pilot might recognise and appreciate as an invitation to a high-quality fight. Just before time 3 in figure 3 the black fighter waits for the grey fighter to unload and get ready for the shot. As the grey fighter draws near at his 4-5 o'clock, our black fellow rolls toward the attacker, and pitches up by pulling toward him as he goes. As the two tracks intersect, perpendicularly or nearly parallel depending on how many angles the defender relinquishes, the defender is hanging inverted over the attacker, canopy to canopy. The black fighter sees his enemy zip by below with his missed shot and calmly finishes his attack by rolling down into the white fighter's astern for the killshot. It all happens in one sweeping motion, without hesitation and without stopping to verify – just pitch up, roll over and down. Do note though that it's nigh impossible to perform the rolling scissor if you're all out of energy – you absolutely must have speed enough to produce a decent rate through the enemy gunsight and speed enough to execute the pitch-up with rate to spare.

If the bandit is careless or slow enough to let you pass by him in his cold six, i.e. out of sight below his belly, it's an easy matter to kill his situational awareness by reversing out of his sight and going vertical. Whichever course of action you decide on it's likely to bring immediate results since you're manoeuvring while he's flying essentially straight, even if it's only for a few fractions of a second.

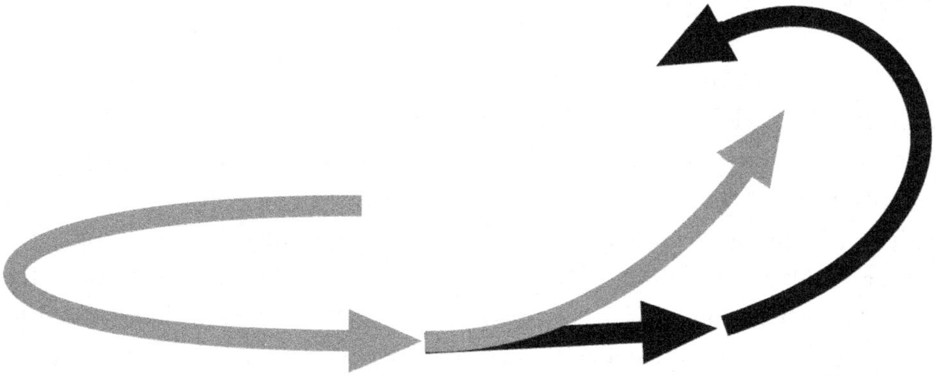

Fig 4. Converting from a static to a dynamic fight

In figure 4 above the two fighters are in a static flat nose-to-tail circle again. The black fighter decides to leave the unproductive contest and relaxes his back pressure on the stick momentarily, thereby going into lag pursuit (i.e. putting his lift vector behind the bandit instead of straight at him). At that moment he starts building energy relative to the enemy, though in the process relinquishing position and angles: he's becoming energy-rich but angles-poor. The bandit, seeing his growing angular advantage, is highly likely to continue hauling back for all he's worth and thereby blowing energy at a sustained rate, because he can see that it's (seemingly) producing results. The black fighter on the other hand, is now free to use his small energy advantage to go into a short high yo-yo, which the grey fighter will most likely counter by cutting across the curve and seek a snapshot. The black fighter defeats the shot by maintaining sufficient manoeuvre speed which allows him to produce enough rate through grey's gunsight, with a particularly tricky oblique vector on the way down from the yo-yo's apex. Provided he survives the shot opportunity the black fighter goes down to "dip into the bowl" for a low yo-yo while the grey bandit is busy recovering behind him, largely bereft of energy after his missed shot.

The black fighter is now converting the static round-and-round fight to a vertical contest. After the low yo-yo he'll be back on the way up again for another high yo-yo, meanwhile the bandit has just recovered from a near-stall and is in no position to keep up. The bandit is stuck in a flat circle whereas the black fighter is describing an oblique circle and turning it into a fully vertical one. Before long he'll be able to haul down into the white fighter from above, or slice into him from a high-side position, and subject him to a series of snapshots. The black fighter is stashing up on energy while the bandit is becoming increasingly defensive and still blowing chunks of energy.

It should now be fairly obvious that it's counterproductive, in any fight, to haul back on the stick for all you're worth as a rote behaviour. Use your brain instead of your muscles, and trust the enemy to make the mistake of fighting reflexively. Countless are the times that I've been assailed by a vastly energy-superior enemy and finished the fight by subjecting the totally out-energied and helplessly flopping bandit to a series of slashing attacks, and that in a crate that was decidedly slower.

The neutral state

In order to fight without fear you need to know when you and your enemy are neutral relative to each other. The neutral state is an "equal opportunity" state: either of you can get into a stern position or extend safely away. You are neutral when:

- You are coming head on toward each other.
- You are going tail to tail (as in having blown through a merge without turning).

- When you are parallel to each other, as in belly-to-belly, canopy-to-canopy or wingtip-to-wingtip.

The neutral state is of course fleeting. Use it as a reference as to whom is earning angles and position in the fight.

CHAPTER 7: LIFT VECTOR, GRAVITY LOADS AND OTHER TERMS OF COMBAT

You have probably heard the hoary maxim of "keep your lift vector on the enemy". What does it really mean? Well, the "lift vector" is always pointed straight up your canopy, so it's only significant in a turning fight where the opponents are actively trying to get their guns on each other. It's a term for describing position and type of pursuit (lead, pure, lag): if you're in a standard nose-to-tail fight with the enemy straight across the circle, you both have your lift vectors on each other as long as you keep honking that stick right back. If neither of you are gaining angles, you're both essentially in pure pursuit, and essentially neutral (i.e. no one has the advantage). Now, if you were to make a half-roll and continue your turn in the other direction, your lift vector would momentarily point diametrically away – until you complete the half-circle and meet the opponent head on. After the merge, your natural instinct is to honk back on the stick and swing onto your opponent's six, right? That is keeping your lift vector on the enemy.

This "lift vector" is significant inasmuch you can actually see how much the opposing fighter is able to pull in a given situation. Say that you meet him head on again after having reversed the nose-to-tail fight to a nose-to-nose fight: if your crate is the better performer, you may be able to fly "outside" (i.e. above) his maximum permissible lift vector (i.e. max permissible Angle of Attack, or outside his flight envelope) and thus cause him to stall out or at the very least defeat his guns solution. Similarly, you can relax back pressure and fly "beneath" his velocity vector, i.e. momentarily going into lag pursuit, forcing the enemy to push negative-G loads should he attempt to get his guns on you.

Fighting is all about knowing what the enemy can and cannot do under given circumstances, of knowing how the enemy is likely to react to given input – and to use this predictability against him. One such dead-certain knowledge is that the enemy will nearly always try to keep his lift vector on you, even if it means burning every ounce of energy or, in the extreme case, ploughing straight into the ground. Consider the case where you're locked in a fight against a bandit with high wing-loading. You, in the better-turning crate, may be able to execute a split-S with several hundred feet to spare, whereas the bandit, reflexively following your move, will auger in dramatically trying to keep his lift vector on you.

Lift and velocity vectors is also at play when you're spiralling or a bandit, or subjecting him to a chandelle (more on this later on). He will most likely try to keep his velocity vector on you to the point of stalling out – because his energy was depleted so much that he couldn't sustain the climb, or rather, because you kept increasing the angles, he exceeded his maximum angle of attack trying to keep his guns on you.

GRAVITY LOADS

Gravity is one of the four central forces you have to contend with, the others being thrust, drag and lift. When you're cruising around, the plane (and the pilot) is subjected to normal gravity – 1G. When you manoeuvre, inertia causes you to experience greater or lesser gravity loads depending on the difference between the velocity vector and the manoeuvre vector. Thus, when you haul back on the stick, you increase the gravity load consonant with the amount of back pressure and relative to your airspeed. The greater the velocity vector the greater the inertia and thus the greater the potential gravity load. If you're really motoring, you don't have to pull a lot to experience very high G's.

"Pulling G's" is a recurring feature of air combat, and one that isn't particularly auspicious. When you pull gravity loads you trade energy for position, or energy for an abrupt change of heading. Pulling sustained gravity loads will bleed your energy to the point of flapping helplessly about, not to mention the very real risk of blacking out, something you should be extremely wary of. Pull less G's in combat and you will be all the more in control and retain all the more energy. To be sure, pulling G's is often necessary: in break turns, when hauling in on a tightly turning enemy, when executing radical manoeuvres etcetera. The trick is to pull those G's wisely!

The most auspicious time to pull G's (or rather, to make a radical change in direction) is when the penalty for doing so is slight: in the vertical going up, when your plane is slowing; when you're at corner velocity and the enemy is above his corner velocity; when the enemy has lost sight. Sometimes though, you really don't have much of a choice. When you need to pull G's to get out of his line of fire or to evade a collision, pull!

You should also learn to watch the enemy for telltale signs of him pulling gravity loads. Identifying his G loads, the approximate force and the timing of them, tells you volumes about his energy state, his experience and his frame of mind. Look especially for wingtip vortices[1] and accelerated stalls, and his behaviour when he appears to lose sight of you, because they spell opportunity.

ANGLE OF ATTACK

I've mentioned AOA a couple of times already. Understanding the concept of AOA helps in realising why you cannot do certain things in the air, such as pulling harder on the stick in an already hard turn. AOA is about the wing shape and its ability to create lift, lift being one of the basic forces pilots have to deal with (the two others being gravity and drag). Lift is of course created by that hard stream of wind carousing over your wings, which in turn is created by the noisy engine up front. The

[1] condensation trails

engine also allows you to gain altitude which you can use in lieu of an engine for purposes of lift, should you experience a sudden loss of motive power.

Anyway, the wind needs to stream at a certain angle and at a certain minimum velocity over the wings to create that critical pressure which sucks the aircraft upwards. If the air stream is disturbed or blanked out be the wing itself or by the fuselage, lift is no longer produced and you fall like a stone thanks to gravity.

The angle of attack is measured between the longitudinal centerline of the thickest part of the wing (called "chord") and the direction of the relative wind, or between the chord line and the flight path. As you surely know, the chord is set at an angle to the fuselage, meaning that though you may be flying "directly ahead" the wing may be at an angle of attack measuring 5 or 6 degrees up in level flight. As you pull up, e.g. for a zoom, the angle of attack increases against the relative wind.

Most importantly, the wings of your aircraft can only create lift up to a certain AOA, usually around 17 degrees. If you exceed max AOA the airflow over the wing will be disrupted and "burble", creating an instant loss of lift. This is known as a stall, and you can experience it at any time by cutting back throttle while keeping your nose high. As speed and thus airflow decreases, you must raise the nose more and more (increase AOA) to remain in level attitude – until such point as you feel a wing drop from loss of lift. Having a crate with a good max AOA at various speeds is beneficial inasmuch as it can sustain a higher instantaneous change of pitch without suffering accelerated stalls, and has a better low-speed handling, such as in low and slow dogfights. Other than that, it's not much to keep in mind. Fly your plane cleanly, don't yank that stick too hard, and you won't have to worry about stalls at all.

ANGLE OFF TAIL

Angle Off Tail, or AOT, is a combat term used to describe pursuit and gunnery situations involving two or more aircraft (also known as "angles"). AOT is never static – it changes fractionally throughout the fight. If you're parked straight behind a bandit, angle off is zero. In gunnery discussion, angle off is normally called "deflection".

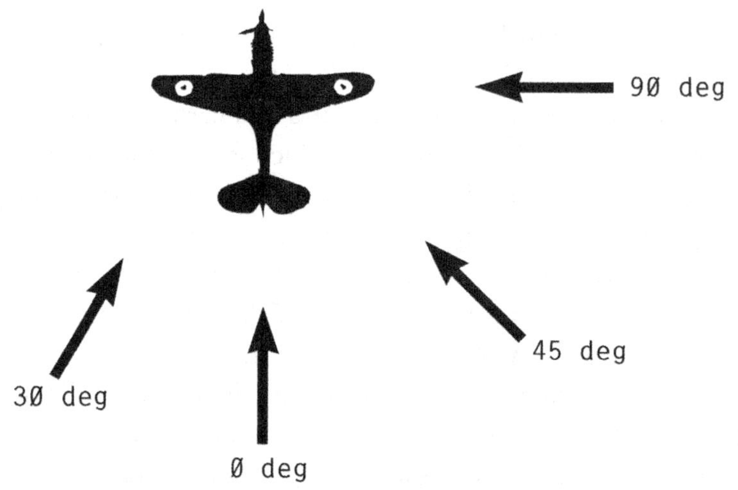

Fig 5. Angle Off Tail. When approaching frontally, angles are used similarly (up to 90 degrees) to avoid confusion, i.e. in a head-on situation, AOT is said to be 0 degrees, although it's technically 180 degrees. E.g. if you're closing in from the bandit's 10 o'clock, you're coming in with 30 degrees of deflection.

Most pilots manoeuvre to place themselves at 0 AOT – they "saddle up" behind the enemy and match airspeeds to gain a comfortable high-probability shot. This is natural and convenient, for it's the position which allows you the most time in the gunsight and poses the least difficulty in aiming. Since the target only rarely cooperates by flying straight and level, unless he's fast asleep, most fights tend to degenerate into turning contests of some kind. Keeping track of angles gained and lost is therefore of some importance if one wants to know whether a particular manoeuvre was successful or not. A high AOT means that the attacking fighter must draw increasingly more lead in guns situations and that the firing window becomes correspondingly smaller.

CORNER VELOCITY

Corner velocity is the speed at which your particular aircraft has its optimum turn performance. You can either consult charts, or get a good handle on your corner velocity through a series of tests. You will notice that your plane doesn't turn very well either at very high or very low speeds, but will perform very well somewhere in between – usually around 280-350 km/h for WWII-vintage aircraft. This is the speed where you get both maximum turn rate (in angles per second) and best turn radius (in meters) at maximum permissible gravity loads: if you're above corner velocity both your rate and radius will suffer, and you are likely to black out from excessive G's and/or suffer an accelerated stall; if you're below corner velocity your rate will be low but your radius may be smaller.

Corner velocity is important to consider when you're jumping someone, or are being jumped yourself by a bandit with a ton of smash. The best course of action in most cases, if you find yourself with an excess of speed, is to shed some of it in the vertical. Thus, if you're being chased around and need to manoeuvre radically, first bleed off some speed by bringing the fight up into the vertical rather than to incur a heavy penalty in the form of blackouts through excessive G's at your present (high) speed.

CLOSURE

The speed differential between two aircraft is called closure. If you're gaining on an enemy, you have [positive] closure. If the enemy is drawing away, you don't have it – your closure is negative. The rate of closure is important to consider, mainly for guns situations, as you will want to close rapidly with the enemy but still have a decent shot opportunity once you get there. In other words, if your closure is too high, your shot window may be too small to deal out a killing volume of fire. Thus, if you're in such a situation, you must manage your rate of closure (or overtake) by shedding some excess speed: by throttling down, by fishtailing or by a vertical displacement. The latter course of action is generally recommended, although when surprise is on your side it may lead to the enemy getting wind of your attack.

Why is closure management important? Well, things will be a lot easier if you can remain behind the enemy – and behind the enemy is anything behind, beside, above or below his wingline out to one turn radius (if the distance is greater, he can turn to face you and you will no longer be behind him, right?). If you're packing an overhead of energy you risk overshooting the bandit before you have time to deal him critical damage, thus you must somehow slow down. It is easiest by far to stash your overhead of speed in the vertical, by zooming up and behind the bandit while rolling to keep him in visual, then to swoop down on him at a more controlled and leisurely pace.

You will want to dispatch your bandit as quickly, safely and effectively as possible, with a minimum of risk and expenditure of precious ammunition, because, the more time you spend dawdling the more of a chance you give him to spot you, to combat you and to get help from his friends. A stealthy, rapid approach to the enemy rear quarter requires a substantial overhead of energy to be used for a lightning-quick attack. By definition this means that your firing window will be quite short, usually no longer than 2-3 seconds. On the other hand, you will want to maximise your firing window so that you have time to stabilize your platform, time to sight your guns properly and time to deliver a lethal volume of fire. So, you will want to approach both fast and slowly! It's not that tricky really: see chapter 17.2 – "The Bounce".

CHAPTER 8: BASIC FLIGHT MANOEUVRES (BFM)

It may seem somewhat redundant to "go through the motions" and describe details of basic flight, but you'd be surprised as to how few online pilots actually know how to land properly or how to perform a simple skid. If you've never ventured into the hostile skies before, this section is highly recommended.

TAXIING

At times, you will need to form up on the runway prior to takeoff. This is a simple enough procedure which ensures that all pilots are facing the same takeoff heading and are properly paired up in wingman teams. To taxi, crank up the engine and knock the throttle forward just a tad to get rolling. You won't need more than perhaps 5-10% of engine power to start rolling. To taxi straight forward, lock your tailwheel (if this is an option). To steer left or right, brake accordingly with your tailwheel unlocked. Make sure to keep your speed low enough to keep the tail down – if you power up too much the tail will rise and you're likely to slew radically to the side in a ground loop. Be gentle with the brakes, or you may find yourself digging holes with the propeller, or even turning over entirely.

TAKE OFF

When you're set to take off, increase engine RPM to maximum, trim the aircraft tail down and open up the throttle, gently at first, while pushing gently forward on the stick to lift the tailplane. This is important, because if you don't your AOA will be too high – the wings won't create lift and you will slither around on the ground in a stalled-out condition. Once the nose comes down and the wings are level with the ground you should be accelerating rapidly. Counteract engine torque with a touch of rudder and possibly a bit of aileron, keep your run straight as you can and let the machine build up to takeoff speed – usually somewhere around 100-150 km/h. Once you reach this speed, let it increase yet further and gently pull the stick toward you. Don't yank it back! A gentle pull is all it takes to become airborne. Once up, crank up the undercarriage and let speed build up yet more before hauling back on the stick. When you're doing more than 150-180 km/h you can safely increase AOA to steepen your climb, or otherwise start manoeuvring.

LANDING

Few sim pilots knows how to land properly. One of the funniest things you can do online is actually to remain on the field and observe takeoffs and landings. It's a hoot. You'll see pilots bouncing and augering left and right, smacking into hangers and groundlooping – don't let it happen to you. The proper way of landing is to let down to a manageable altitude, generally about 500-1000 feet, with your intended runway direction directly to your side. Normally one flies a counter-clockwise leg, meaning that you keep the field to your left. Fly past the runway at this altitude while shedding speed and altitude. As you pass the airfield boundary, continue on for a while and then make a coordinated turn to the left to get onto your glide path (the final stretch, in line with and pointed down onto the runway). As you're about to cross the glide path, make a final turn onto it and line up with the runway.

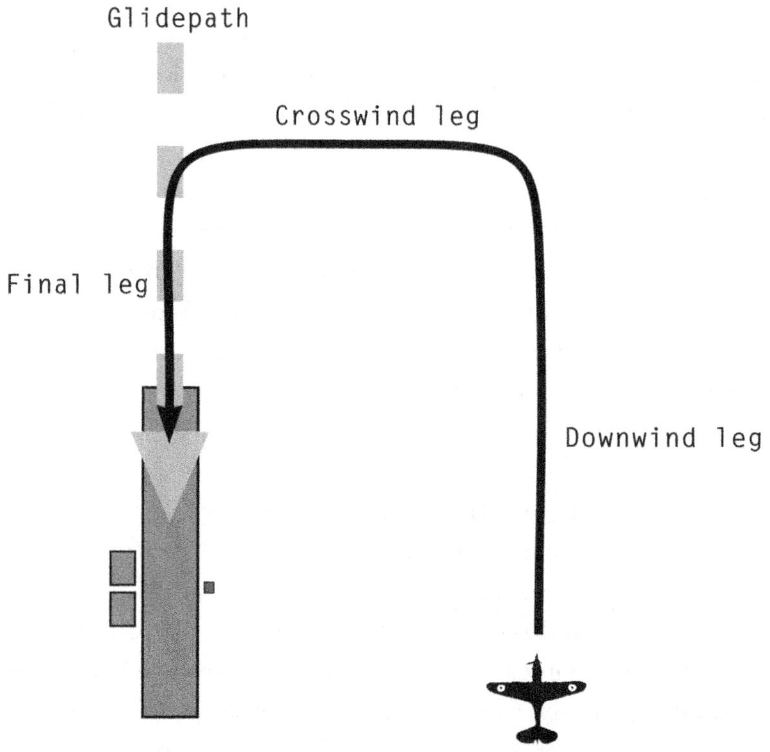

Fig 6. Preliminary landing procedure.

You should now be at about quarter or half engine power, have your flaps out, your landing gear down and have your aircraft trimmed tail heavy. Now, the objective is too keep your nose slightly above the horizon (only so much that you can still see the runway) and *use your throttle only* to adjust the rate of descent and approach to the first quarter of the runway. In a perfect landing, you should not have to adjust your pitch (nose up/down) at all, and should be using no more than 10-20% throttle on the approach. If you notice that your descent is too rapid – that you will most likely land short of the runway – increase throttle to bring you back to the proper glide path. If you're about to overshoot, cut back to idle. If you're coming in both short and hot (too low and too fast), chop throttle and use the extra speed in a short ascent to the glidepath. If you're really overshooting the runway, i.e. going far too fast and far too high to glide down on idle throttle, break off the landing attempt and set up anew.

As you draw closer to the runway at just above stall speed (about 90-110 km/h depending on your crate) with your nose slightly high, you will get a good feeling for how an increase or decrease of throttle affects the glide path. Shortly before touchdown, actually just a few feet over the runway, cut throttle entirely and raise the nose just a tad more – this is known as to "flare" the aircraft – and let the aircraft settle on the ground. The slower you fly, the less you will bounce and the sooner you will come to a stop. Let it roll and pump the brakes gently to bring down your speed. Do this a couple of times until you can make a perfect landing every time, and to get a good feeling for initial altitudes and speeds. When you've become proficient enough, make it a habit to check your VsI indicator (vertical speed gauge) while landing. A descent rate of 1-2000 feet per minute indicates a gentle glide path whereas 3-4000 fpm is clearly excessive and will most likely lead to permanent damage and injury. On final approach, your descent rate should be on the order of 2-400 meters per minute for a nice and gentle touchdown.

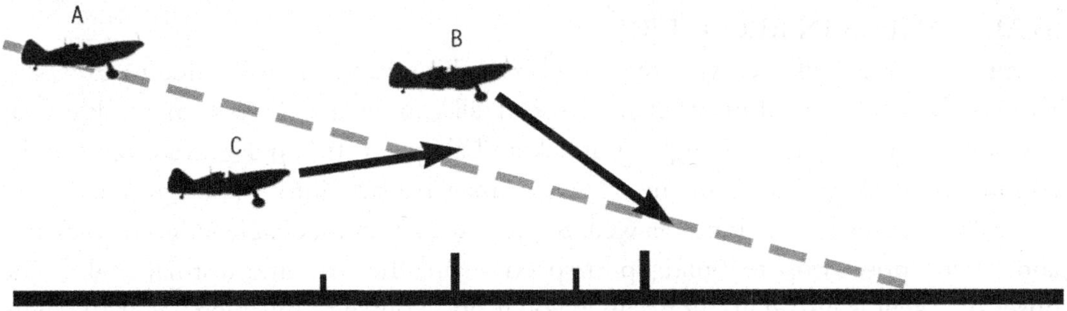

Fig 7. Landing. Manage your descent rate with throttle to stay on the glidepath. Crate A is good to go. Crate B is too high and must chop throttle. Crate C is too low and must throttle up to regain the glidepath.

COORDINATED TURN

A coordinated turn, as opposed to an un-coordinated one, employs a measure of rudder. All flight simulations lets you turn without the use of rudders whereas a real-life pilot would never consider such an abomination due to effects of gravity and parasitic drag. Turning is simple: give a bit of rudder in the turn direction at the same time as you bank the aircraft gently in the turn direction, and introduce just a bit of elevator pressure to avoid losing altitude in the manoeuvre. Once in the turn, neutralise rudder and aileron, and maintain a steady back pressure to sustain the turn. Practice turning into cardinal headings – north, west, southwest and so on, to get a hang of the forces and timings involved.

SKID OR SIDESLIP

The skid is a convenient way of shedding altitude without changing direction, and can also be a part of a combat manoeuvre. The ingredients of a skid or sideslip is to keep the nose high, to give full rudder deflection and to counteract the roll with a bit of opposite aileron. This will increase drag significantly and cause you to rapidly lose altitude and airspeed – imagine doing that with a bandit on your tail: odds are high that the opponent will overshoot (fly past) and end up right in front of your guns. Note though that the skid robs you of significant energy and isn't really advisable other than as a last-ditch manoeuvre, and preferably when you have some altitude to play with already. Practice making aggressive skids by banking 45 degrees to one side and giving full rudder deflection in the opposite direction, nose high – you will drop like a stone. Another way of shedding speed and altitude is to "fishtail": pumping full left and full right rudder alternately with your nose high. The increased drag, caused by the increased AOA and by the fuselage disturbing the airstream, will decrease your airspeed significantly. This is useful when you're coming in hot on landing or when you're at risk of overshooting (i.e. overtaking) an enemy from behind.

STALLS AND SPIN RECOVERY

Getting out of a stall is easy. Since a stall is nothing but loss of lift due to excessive AOA and/or a factor of insufficient airspeed, all you need to do is to neutralise your controls and regain normal angle of attack and airspeed. If you aggravate the stall by continuous back pressure on the stick you may happen into a spin, although that generally requires rudder input as well. So, if you stall, immediately let go of the stick and let the nose drop to build up airspeed again, then resume normal flight. The onset of a stall is also quite noticeable: you'll find yourself "mushing" as if the crate was flying through molasses, and the aircraft will yaw noticeably towards the stalling wing (one wing always stalls before the other) before the wing actually drops from loss of lift. Now, if you're already on the deck you'll have to be very careful about your recovery as it's nigh impossible to recover without a certain loss of altitude. If you maintain back pressure on the stick in a stalled out condition, you will quickly lose all forward speed and end up in a spin.

When the aircraft spins, it gyrates around its center of gravity (CG) while falling more or less vertically. This is because the wings no longer produce lift, because there is no forward motion. Should you happen into a spin, the proper procedure is to neutralise controls and throttle back, then input rudder opposite to the direction of spin. I.e. if you're spinning to the left, let go of the stick (just release your grip and let the stick center itself!) and give full right rudder. As soon as you recover from the spin, neutralise controls again (stop kicking the rudders!), throttle up and let airspeed build up before trying to resume normal flight, lest you enter a spin in the other direction. It is always best to try and get out of the spin as soon as it begins – the longer you wait the higher the AOA and the more difficult the recovery. Practice getting in and out of spins at various altitudes. Spins are easily entered by exceeding AOA with full rudder deflection, preferably at low airspeeds.

AILERON ROLL

Fig 8. Aileron roll.

In level flight, roll the aircraft swiftly through 360 degrees by holding the stick fully to either side until you're "sunny side up" again. As you complete the roll, stop the roll by moving the stick rapidly in the opposite direction to recover level flight. You should come out of the roll at the same altitude and in the same direction of travel as you began it. You may also have heard of the "snap roll", or involuntarily experienced it. The snap roll is actually a stall manoeuvre in which one wing momentarily loses lift, aggravating roll rate. It's easy enough to perform, but be advised it burns a bit of energy and may leave your situational awareness in the pits: move the stick fully to the side and then pull full back in a single deft swipe. This will stall the low wing and "snap" it, increasing your roll rate. Once you've snapped, neutralise controls to regain lift and controlled flight. You can sustain the snap roll at a significant energy cost, or snap and then continue with a normal aileron roll.

You should also practice quarter-rolls and half-rolls. Make your moves determinedly and precise. Practice doing a quarter-roll – hold – roll to inverted – hold – another quarter-roll and hold, then back to upright. Same thing with the half-roll, roll to inverted and hold, then either continue the roll or roll right back in the opposite direction. Measure your success against your ability to maintain your general heading and altitude throughout the manoeuvre.

BARREL ROLL

Picture yourself as a pea rolling forward and around the inside of a great barrel lying on the ground. This is a tricky manoeuvre as you need to balance aileron and elevator input to maintain your general heading. You should pull back gently while rolling, but not so far as to exceed your maximum Angle of Attack (the point where lift is no longer produced) or so much as to cause a blackout. Also note that the amplitude of your back pressure varies from considerable to minimal at various points through the manoeuvre, i.e. your elevator input should not be constant but variable. You should come out of the roll at the same altitude and in the same direction of travel as you began it. While disorientating at first, it is actually a rather delightful and graceful manoeuvre to execute once you get used to it. It helps somewhat to keep your eyes fixed in the forward-up view, maintaining a grip on the horizon, so as not to end up in an oblique circle or flushed out perpendicularly to the intended heading.

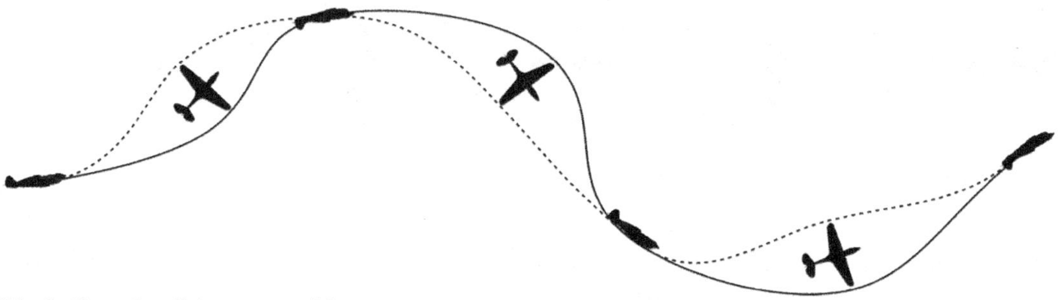

Fig 9. Barrel roll (not to scale).

The barrel roll is a great manoeuvre for transitioning from defence to offense against a bandit closing in to guns range in your stern quarter – it slows your forward progress without an altogether crippling expenditure of energy. If the enemy is concentrated on a largely unloaded guns run, you will "corkscrew" along his flight path, producing an overshoot which you capitalize on by sliding in on his six. The barrel roll is not restricted to a single 360 degree rolling manoeuvre but can be sustained through as many revolutions as you wish (or have energy for). Also note that the barrel roll is useful in a head-on engagement, where it's employed as the first part of a lead turn manoeuvre. More on this in later chapters.

LOOP-DE-LOOP

A loop is a simple aerobatic manoeuvre in the vertical plane. Pull back and maintain a gentle back pressure. Watch the horizon in your up-view, and adjust any deviation from the true heading with a touch of rudder. Exit the loop at the entry level, and in the entry direction. The loop is not a combat manoeuvre. I'll say this again: the loop is NOT a combat manoeuvre, though you will see lots and lots of pilots using it as one. The loop is one of *the* most predictable manoeuvres you can pull in a fight, because it's so very time-consuming, as it allows the discerning enemy plenty of opportunity to cut across your flightpath and gun you down, usually at or near the top of the loop.

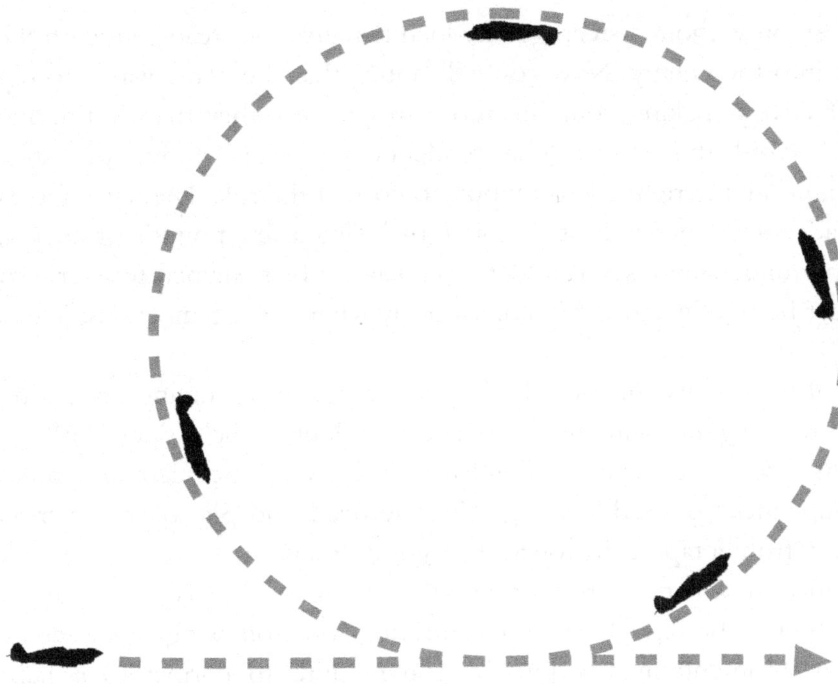

Fig 10. Looping is nice, but it's not a combat manoeuvre.

REGARDING RUDDERS

Proper use of rudders is something which takes quite a lot of practice to master, mainly because the lack of sensory input in computer flight. Use of rudders is not vital – because the physics model isn't sophisticated enough (no wind gusts, no prop wash effects) and because the plane can be sufficiently well flown without them. That said, when you do get the hang of rudder play, you will never want to fly without it, and will return to base immediately if it's shot to tatters. Flying without rudders is like driving a car on the rims.

The stick controls pitch and roll. The rudder controls the third plane: yaw, or the left-right slew off the longitudinal flight path. Its function is quite similar to that of turning the steering wheel of a car, though with the difference that the surface isn't solid. In other words, the plane will exhibit a rolling tendency in the direction of yaw, and you will incur an increasing drag penalty for as long as the rudder is in use. This can be both a benefit and a disadvantage depending on the circumstances. When you're giving a lot of rudder you burn a lot of energy in exchange for a somewhat better turn rate. On the other hand, the yaw will throw you off kilter with regard to matching the enemy plane of manoeuvre unless you counteract the resulting "over-rolling" tendency by giving a bit of opposite aileron. E.g. in a 45 degree banking turn to left at corner velocity, you might want to seek a lead pursuit position on the bandit across your circle. Since a sustained elevators-only turn will not yield such a position without a steep gravity load penalty, you tread gently on the left rudder to yaw into the enemy. Now you will notice that the plane wants to dip into the circle, effectively making your life more miserable rather than improving your position. To avoid this over-rolling tendency you must move the stick somewhat to the right (giving right aileron input) to correct the roll. The net effect is that your angular advantage increases at the price of losing a few pounds of energy. As you will no doubt understand, such rudder input cannot be sustained forever – to the contrary, it must be briefly and deftly applied only when it's the most auspicious to do so.

Rudders are also most useful for side-slipping, fishtailing, standing on a wing without loss of altitude (e.g. left wing down, full right rudder) – the rudder works as your elevator in this case – and for correcting (or walking) your fire. Certain combat manoeuvres are dependent on rudders, e.g. Hammerhead and Sliceback, whereas others benefit greatly from a tap of the toe at the right moment.

Whenever you have a bandit on your six, in guns range, the use of rudder is most auspicious. Since the enemy is concentrating on following your every manoeuvre you can momentarily fox him by forcing him to correct to a false heading. A small skid, yawing the aircraft to one side, will lead the enemy to pull deflection in that direction while you're actually travelling in the opposite.

Fig 11. Skidding to slow down and to trick the enemy into drawing false lead (not to scale).

In the picture above the defending fighter is skidding by giving full rudder to one side while compensating the rolling tendency with opposite aileron. This gives the impression that he's heading either right or left (thin arrows) while in reality his general heading is largely constant (fat arrow). Drawing lead against such a target is extremely difficult since it defies everything you've learnt about deflection shooting – perception does not equal reality! The pursuing fighter is led to compensate for this without the use of rudder, forcing him to use ailerons and elevators to stick with the perceived manoeuvres (dashed line). Since use of rudder increases drag, the defending fighter slows down considerably whereas the pursuer largely maintains his speed, eventually leading to an overshoot. Do note however that this ruse only works for a limited time and does expose the defending fighter to devastating fire from an enemy who is able to accurately assess the manoeuvre.

The best way of learning how to use rudders is through trial and error. If you have them, use them! Don't be afraid to experiment – at worst all you have to do is replane. Practice using rudders in turning fights: use top rudder to skid up, down rudder to slice in. A smidgeon of rudder usually earns you the few angles necessary to get your guns on a predictable target. In a nose-to-nose fight in particular, you'll find that you can fly on the inside of the bandit (i.e. outside his performance envelope) and skid outward with rudder to ream him on every near-head-on pass. Similarly, when a bandit breaks hard and you're behind him at or near corner velocity, you can yaw your shot in – or skid up preparatory to slicing down into him. Thus you will make seemingly "impossible" shots against an enemy who doesn't realise the utility of rudders. I remember when I first got my rudder pedals: I hardly dared tramp on them for fear of what might happen! When I finally learned how to use them, I never looked back.

CHAPTER 9: PURSUIT MODES

Whenever you're manoeuvring against an enemy or formatting on a friendly aircraft, you will have to employ one of three possible modes of pursuit: lead, pure or lag pursuit. These modes of pursuit signify your heading (or rather, velocity vector) relative to the target. They also govern your angle off tail to the target, and your closure rate. In most fights you will generally switch seamlessly between these three modes. Knowing when to employ a certain type of pursuit, and seeing which kind of pursuit the enemy employs against you, is a central aspect of air to air combat. We'll be using these terms again and again in discussing various manoeuvres, so make yourself comfortable with them. As a beginner pilot, it may help to actually enunciate the various pursuits you're employing at the time. Don't be afraid to talk out loud, after all, you're probably sitting all alone in your dungeon anyway so who's to notice?

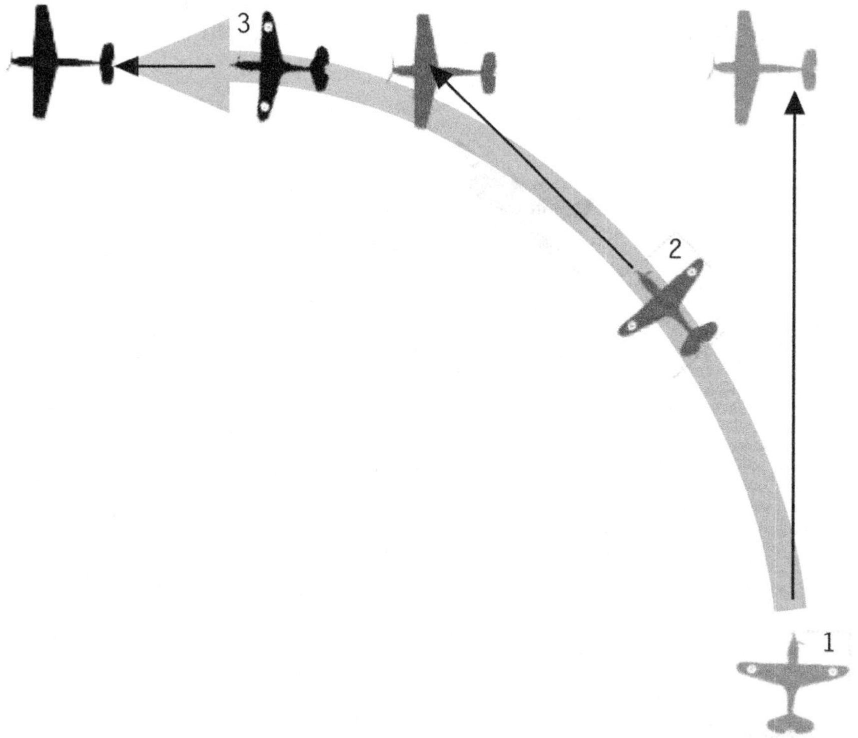

Fig 12. Pure pursuit. For those times when you really want to saddle up.

PURE PURSUIT

Whenever your velocity vector (in normal flight this is signified by your gunsight pipper) is pointed straight at the enemy, you're in pure pursuit. If your rate of closure isn't excessive you will eventually end up directly behind the enemy. However, if the target is turning, angle off tail is increasing all the while, thus creating a high angle off guns situation. Since the target is always moving at a certain velocity, the pure pursuit curve isn't ruler straight as one might think, but slightly curved, as can be seen in figure 11 above. As you can see, AOT decreases from challenging (90 degrees) at time 1 over tough (45 degrees) at time 2 to dead simple (0 degrees) at time 3. While the easiest and most natural of pursuit modes, pure pursuit introduces a slightly more challenging shot (since you're pulling G's, however little) until you draw in to dead astern. Pure pursuit is also more time-consuming than a lead pursuit curve, and thus more predictable. As you ride the stall in a standard balls-out max rate turnfight, straining to keep behind the bandit, you will be essentially in pure pursuit, following in his wake as it were, even though you may not be actually pointing your nose at the bandit. It is also plain to see in figure 11 that if the attacking fighter were to fire either at time 1 or time 2, his fire would fall ineffectively well behind the bandit. This is why you must normally "pull lead [pursuit]" in order to hurt the badguy, which takes us to the next mode:

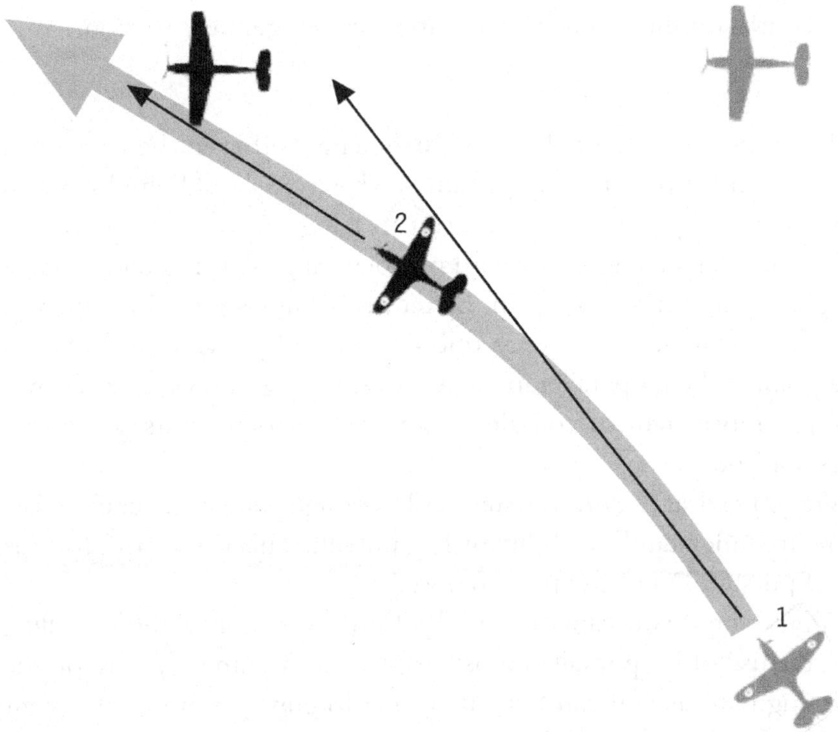

Fig 13. Lead Pursuit. Quicker than pure pursuit, deadlier by far.

LEAD PURSUIT

Whenever your velocity vector is pointed ahead of the enemy, in his future flight path, you're employing lead pursuit. Think of it as a way of cutting off the enemy, either in the horizontal or the vertical plane of manoeuvre. As can be seen in figure 12, the attacking fighter points his nose well forward of the bandit at time 1, in his projected flightpath. Provided that both aircraft flies a steady course, AOT is constant. At time 2 the brave attacking fighter makes a quick jink, a small correction, to adjust for the target's speed and lets off a nice long steady burst just in front of the enemy. He keeps up his fire until the enemy has flown fully through the sheet of flame, who promptly explodes.

The employment of lead pursuit is the only way a slower fighter can catch a faster target. In the bulk of combats, the attacking fighter either maintains lead pursuit throughout the attack and goes for a brief high angle off snapshot followed by a reversal in the bandit's rear hemisphere, or, by switching to pure and lag pursuit as required when he draws into range, converts his position from high angle off to low angle off tail and parks on the victim's six for an extended gunnery session.

LAG PURSUIT

Whenever your velocity vector is pointed behind the enemy (either in the horizontal or vertical plane), you're in lag pursuit. Lag pursuit is what you should employ when your rate of closure is excessive, or when you want to hang on to your energy in a situation where the enemy is bleeding his by hard manoeuvring. This is a comfortable way of reducing angle off tail for a subsequent better probability shot, and helps you to remain in the enemy's rear hemisphere, as opposed to lead pursuit which carries you forward to the point of being parallel in some cases. Going into lag pursuit may even cause the bandit to lose sight of you altogether, as it generally puts you where he cannot easily observe you.

Most pilots grasp Lead and Pure pursuit easily enough, but it generally takes some effort and time to understand the utility of Lag pursuit. This is too bad, for it is a most useful kind of pursuit. Do please pay attention!

In a turning fight where you cannot draw lead on the bandit without stalling out or blacking out, the use of lag pursuit is most auspicious. Assuming you're on the bandit's tail and turning hard as you can but cannot get to guns position, relax your turn to slide into his cold six, or to slide up in his high quarter rear for a yo-yo, depending on your vertical attitude. From that moment on you're stashing up on energy but losing angles. It is a rather small matter to convert this energy to angles or a snapshot opportunity again, either on the top of a vertical manoeuvre where you get a gravity assist, or at the top of your small high yo-yo.

By using lag you're displacing your own circle relative to the enemy's while remaining behind his wingline and in control of the situation. The bonus here is that it goes against the grain for the defending pilot to do anything other than keeping his lift vector on you (i.e. honking back hard on the stick) – which in most cases means that he will be turning as hard as he can, thereby burning his energy while you're conserving yours and even building up to a significant energy advantage.

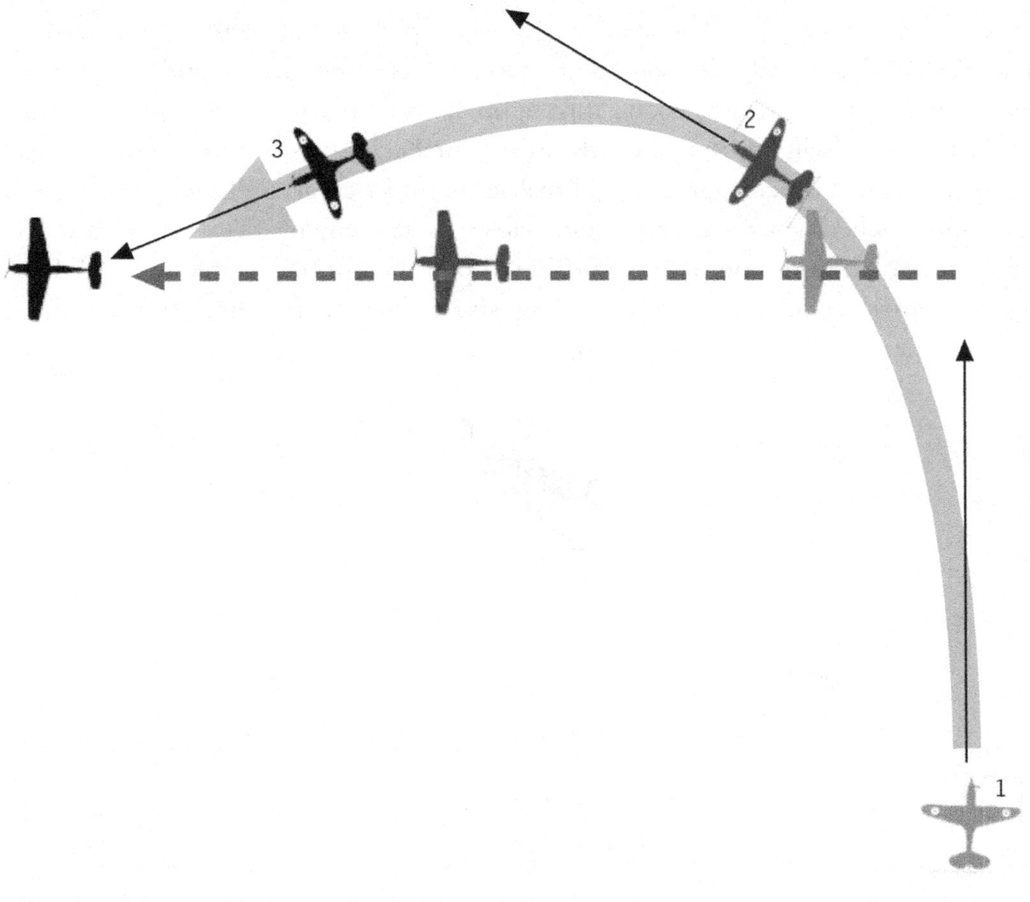

Fig 14. Lag pursuit allows you to remain behind the enemy. At time 1 the attacker resists the urge to point directly at the bandit on account of his radical closing speed, allowing the target to fly across his view instead. At time 2 he is curving in behind the bandit, still in lag pursuit, and shedding energy gently in the vertical. At time 3 the attacker is heading down into the enemy, having converted to pure pursuit and should in fact be settling down to a slight lead pursuit.

When you use lag in a turning fight, depending on the distances involved, you may have the chance to "hide" below the enemy in his "cold six" where his vision is obstructed by his own rear fuselage. Odds are that the bandit, unless he's an old fox, will realise that he's lost visual contact and may *relax his turn* rather than manoeuvre radically to regain visual on you. If you see the bandit straighten out or continue a benign turn, then you can be fairly certain that he's lost you, allowing you an easy shot from dead six. Out of sight, out of mind! And pretty soon, out of the sky too!

In the case described above, you might have started out in pure pursuit (essentially flying in the bandit's wake if you were locked in a static flat circle) and unable to pull into the lead pursuit required for a good guns position, because it would have caused you to stall out and crash horribly to death. By relaxing back pressure on the stick you automatically go into lag pursuit and increase your airspeed (because drag decreases), allowing you to momentarily draw below and closer to the enemy. However, the bandit is still turning as hard as before while you are turning less, therefore he will eventually climb onto your back unless you can hide yourself quickly! From this description you will realise that it's best to pull into lag pursuit in this manner when you're already quite close to the enemy. You may know the feeling: you're on his six but juuuust unable to pull the last few degrees of lead. Relax elevator pressure and scoot into his low six, recover gently and then use your recouped energy to gain the shot position.

CHAPTER 10: SITUATIONAL AWARENESS

"I was at 3K at 400 so my E was still good. I did a quick check of my six and saw no immediate threat. I proceeded to perform a chandelle to bring my heading to the south. My thinking was I needed to get a visual on that 109 before I could know how to extend from it."

Jeff "Gorian" Gonzales, 56th FG(v)

Situational Awareness, or SA, is a catch-all term for the real-time ability to acquire and process a host of different data in a constantly shifting environment, and the ability to translate an assessment into action aimed at maintaining integrity (of self, of dependants, of mission). In short, it means "knowing what goes on, and successfully adapting to it". SA is what makes or breaks a combatant – it's not the vehicle, not the numbers, nor the situation itself. You've heard the saying "it's the one you don't see that kills you" and that is very much so. If you don't see the threat, if you're not aware of it, then you cannot manoeuvre profitably against it. Lose sight, lose the fight!

"Knowing what goes on" is a complex, very taxing and constantly ongoing process because you must identify and keep track of everything that goes on around you in order to make the correct tactical decision at any one time, all the time. You must do it without pause, and the input changes constantly. Good SA starts with, but is not limited to, looking around you, and for that you need to develop a routine. Your scanning must be continuous and cover ALL parts of the sky. Work either clockwise or counter-clockwise, whichever feels best, and look properly – not perfunctorily. To check your six, quarter-roll and jink while checking high and low astern – both left and rear. When you've gone through all the views, start over again, and keep doing it until you land and shut off your engine. Realise that "continuous" means just that: you should be checking every last piece of the sky every 3-4 seconds. Why? Because it doesn't take more than a couple of seconds for someone to drop in on your six, that's why!

Fig 15. The grey cone is your blind area – roll and bank to observe it every 4-5 seconds

A large proportion of pilots see far too little, and understand even less, of what goes on around them. This failure to recognise and react to threats is what gets them killed, not their ability or lack of ability to manoeuvre. Scanning and seeing is the first part of SA, for the rest it is threat assessment based on position, energy and vectors, a matter of prioritizing and storing bits of data in your memory banks to be checked upon again later, and it's a good part anticipation and educated guesswork.

This other part is what one might call "wide sky awareness" and relates to what you might reasonably expect in this part of the world. This ties in to the "big energy picture" discussed earlier – what is the general bogey energy level around here; at what altitudes are you likely to find the enemy; what are their missions, configurations and general headings; will the enemy be on his toes or will they be dreaming on a lazy climb to altitude; are there likely to be massive furballs in the area; will you find bombers and escorts at altitude; is this predominantly a cruise-through area; how likely are you to find friendlies around here etcetera. All these factors, coupled with your own flight profile – are you in the weeds or at superior altitude? Are you cruising at reduced RPM or making best time? Is your aircraft conspicuous or not? – should weigh in to affect your scanning procedure and readiness for action. One thing you can be certain of though, is that there is ALWAYS someone who has more smash and more altitude than you. You can never ever let your guard down. Ever.

Fig 16. Is this you? A lot of pilots fail to look about them properly. Know that it doesn't take more than a few seconds for any sector of the sky to "grey out".

The third part of SA is psychology and anticipation. Let's say that you bounce a guy but miss your shot and use your superior energy to disengage. You see the bandit come after you for a while until he eventually breaks off pursuit and vanishes in the haze. What can you reasonably expect from him? Did you make him mad enough to come after you or is he likely to continue his mission without further ado? Will he climb beyond visual range and loiter in the bounce area, waiting for you to come back? Is he a dogged fellow or one that loses heart easily? Did you give him reason to believe that you might come back or might he assume that you're disengaging for good? Are there other bogies in the area that might catch his attention? Only you can tell, and you can tell only by looking at his moves, by making assumptions from his observed behaviour.

The situation is always in a flux. Throw in a few variables and it changes completely. What if you have a wingman, or a friendly of some sort and ability, to draw the bandit's attention? At what point can you safely reverse? Should you drag for the friendly or can the friendly drag for you? Will the friendly blow his position and become static? What if you're on the deck, is AA a factor for either of you? Can you use the terrain to mask your disengagement or reversal? Is the bogey you saw earlier on a friendly or a bandit, and what is his SA and intent? Is your current heading taking you further into a high-energy bandit-infested area, or out of it? Such questions and deliberations must run through your mind without pause.

Playing this game is such a gamble you'd think you'd learn from your mistakes but every once and a while it happens. Just a reminder telling you not to enjoy it too much. Forgetting about the sky above you or what happens after you've gone in. Wingman with you and yet you don't know where he is. You haven't checked. It doesn't matter if you're both playing element leader tonight. One bandit down, you climb in view of the airfield and you're trying to figure friend from foe. Wasn't there three of us? Who's the fourth guy then? Others have no visual until it's too late. 109. Great, co-altitude right near the field. Lead turning him now and on his six to the south. He knows his game and drops down in the flight path. Options? No chance of catching him but maybe the others have. Disengage? No wait it out a bit and see if friend can catch him. No joy. Conscious of two other cons below you with a friendly above them going in you follow the 109 in a chandelle. He comes down easily evaded. Friendly collides with two below you which are now quickly joining the fray. Still on the first 109 who is now dragging for the two behind you. Smart. Defensive now and time to disengage east. Dropping down for some speed and scissoring you lose one but others stick to you like glue. Failed downward spiral and feeling pretty stupid they bag you in another scissoring run on the deck. Where did it go wrong?

Hammered, 56th FG(v)

In a combat situation, the amount of data is often overwhelming and sometimes conflicting. The normal reaction to that is to deal only with the most obvious or the most critical information while disregarding everything else. Problem is, the things you disregard are highly likely to come after you and bite you in the bum. Literally. The less SA a combatant has, the easier it is to defeat him. Some guys doesn't have any SA at all – these are the easiest marks of them all. Others may have a reasonable SA from the outset, but become distracted by a lesser threat or suffer from tunnel vision. These guys are also easy marks, and they stand out like a sore thumb in the sky. Still, even the most developed SA can be overtaxed and knowing how to produce such a situation is paramount for prevailing against an "ace".

Fig 17. Tunnel vision. You cannot fight what you cannot see.

When you concentrate on a single enemy, as in a lengthy tail-chase or when looking through the gunsight, your situational awareness is in the pits. Keep up your scanning routine at all times, particularly when you're getting fixated on a bandit and when you're about to fire. As a beginner, it takes all your powers of concentration to focus on the bandit up front, your closure rate and your guns solution. This high level of concentration may be required of you for several tens of seconds, even minutes on end if the bandit is un-cooperative. This concentration on a single target at the expense of all else is lethal, as you have surely found out many a time. It does take a supreme effort to rip your eyes off the target and perform your standard scan – know though that it may suffice to make an abbreviated and rapid scan: Check high right rear. Front. High left rear. Front. Adjust. Low right rear. Front. Guns guns!

LEVELS OF SA AND TASK OVERLOAD

In order to progress up the SA ladder, you must first master your aircraft and grasp the fundamental concept of Energy. Having a bunch of bandits at your six is not a problem if you KNOW that they don't have any closure (i.e. are not closing to guns range) and KNOW that if you just keep trucking you'll eventually lose them. A rookie pilot however is quite likely to make the poorest possible choice in that situation - such as burning his energy by excessive turning or panicking by diving out. Here then, is a general table of SA levels:

0. The pilot cannot perform any action with any degree of success.
1. The pilot is able to maintain coordinated flight, read his instruments and track one other aircraft.
2. The pilot can keep track of one other aircraft without looking at his instruments or visual references, regardless of attitude.
3. The pilot can keep track of several aircraft without effort and navigate at the same time.
4. The pilot can keep track of all aircraft in his vicinity and assess their energy state in relation to his own at all times.
5. The pilot can anticipate threats beyond visual range.

This last point is not as cuckoo as it sounds. Upon asked how he managed to pick up enemy aircraft well before anyone else did, famous RAF pilot George Beurling replied "I can smell them". So, whenever you feel the hairs rising in the bottom of your neck, or save yourself by a break turn based on a hunch, you know that you've attained good SA. This also is a factor of having a good grip on the "big energy" picture, indicative of having a good "wide area SA".

The fewer things you have to worry about, the higher your SA potential. Assuming the pilot has a basic SA level (3), what improves or destroys his SA is the number of tasks he has to perform, or the amount of data he has to process. It is also a matter of training and discipline: target fixation or the employment of inferior formations for instance, can be eliminated through training. However, introduce sufficiently many new or diverse tasks to the pilot, and he will become overloaded to the point of not having enough SA to survive very long. He will lose control of the situation and fail to recognize a critical threat. Imposing task overload on the enemy and retaining control over your own tasks is what SA is all about.

What are the tasks? Beyond avoiding an unfortunate contact with the ground, the tasks include:

- Navigation (staying in the air, plotting current and projected position)
- Systems checks (fuel, engine temperature, system failures etc)
- Ongoing visual checks of the surrounding airspace
- Radio communication
- Adherence to mission
- Position and energy state relative to friendly aircraft and territory
- Position and energy state relative to enemy aircraft and territory
- Position and energy state relative to unseen but likely enemy aircraft

For a beginner the first few tasks are sufficient to soak up the bulk of his attention. Progressing down the list the big unknowns are of course the enemy inputs. These are exponential: one bandit (or a formation of bandits) may cause you to devote 20% of your attention, whereas two cons with vastly different data (altitude, speed, attitude, type etc) may force you to devote 100% of your processing effort to sustain yourself. Introduce one more unknown set of data or make one of the previous critical (such as losing a vital control surface) and presto - task overload. So, what separates the rookie from the ace is not really flying ability, but the ability to process and adapt to huge amounts of data.

How do you avoid task overload? Well, one of the most fundamental solutions is to fly in formation. Formations, beyond the basic hypothesis of providing a numerical superiority, help to share the tasks and thus lessen the probability of task overload. Formations add more eyeballs and brains to the scanning and threat assessment routine. However, formations introduce another set of tasks that can be equally devastating (keeping station, responding to orders, keeping on top of the radio chatter, responding to calls for help etc). Therefore, the understanding and employment of appropriate formations is crucial - not only for the individual but for the formation as a whole.

There are other ways, beyond constant scanning, of keeping your SA high and most of them are quite self-explanatory. That said, you'll notice that 99% of all online pilots disregard them 99% of the time:

- Maintain superior altitude and speed.
- Limit yourself to single passes on the enemy, after which you promptly disengage to regain SA.
- Don't surround yourself with enemy (as in a furball).

I said earlier that imposing task overload is key for defeating ace pilots – it is naturally also true for defeating anyone. How do you do that? It's fairly easy actually: as long as you fight predictably, e.g. keep flying a simple nose-to-tail circle with the enemy or stick to an entirely predictable boom&zoom attack, you establish a pattern which the enemy gets to expect and can plan against with little effort. If you, on the other hand, keep changing your position relative to the enemy and force him to devote a lot of energy just to keep track of you, such as in a rolling scissors or by subjecting him to attacks from always changing quarters, then you drain his SA. Add a couple or more friendlies to the equation, all attacking from different angles, speeds and times, and the ace SA will be sorely taxed.

How do you know when someone's SA is not up to snuff? That too is pretty easy to establish. When you bounce someone and manage to motor up on them from a highly visible position, sometimes through a veritable cloud of enemy, and they still don't manoeuvre, then you know his SA is less than stellar. When you fight a guy and you see him make a completely false turn, then you know he's lost sight of you. I've actually flown up to format on the wings of enemy aircraft more than once, and remained in perfect formation with them for minutes on end, without them noticing. You get a hearty belly laugh when they finally see you and make a wild turn away! You can also tell from the way people arrange themselves in "conga lines" – enemy and friendly aircraft interspersed in a long tail-chase – that their SA is somewhat lacking. This latter phenomenon is surely a blend of bad SA, target fixation and wishful thinking, neither of which is particularly conducive to a long and healthy (virtual) life. Bad SA is quite frequent near airfields – for some reason people consider themselves safe near a friendly base and seldom check either their six or the immediate surroundings. Feast upon them, and make sure you're not the ready-made dinner yourself.

Enemy fighterbombers are also extremely SA-challenged, particularly so when engaged in CAS. If you happen onto an enemy presence low over a contested locality, you can be pretty certain that they are observing the ground first and the sky second. Telltale signs include tracer going down, bomb-bursts, bogies seen flying Cuban-8's and Hammerhead's over the target area and a complete disregard of your sudden appearance. Choose your target wisely (i.e. attack him while he's seen to busy himself with ground observation or during an extension from a strafing run) and make your attack so as to avoid reversing anywhere near other bandits in the area. You will want your mark to be flying reasonably straight, and the best opportunities are on his run in to the target and shortly after it.

WHAT IS A GOOD FIGHTER PILOT MADE OF?

"A fighter pilot must possess an inner urge for combat. The will at all times to be offensive will develop into his own tactics. I stay with an enemy until either he's destroyed, I'm out of ammunition, he evades into the clouds, or I'm too low on gasoline to continue the combat."

Colonel Hubert "Hub" Zemke, 56th FG

Beyond having a working situational awareness, keen eyesight, a stable mental state and a fair dose of controlled aggression, the above-average fighter pilot must have extraordinary reflexes and the ability to anticipate the moves of his enemy – both the individual enemy and the overall threat picture. He must be able to discipline himself: to prevent himself from making potentially lethal mistakes; to recognise dangerous situations for what they truly are and act accordingly; to avoid becoming greedy. He must be fully conversant with his own and all enemy machines and their envelopes under all circumstances to fully appreciate which manoeuvres are possible and which, in conjunction with the perceived enemy mental state, are likely in any given situation. He must feel that the machine is a living extension of his body, not a strange vehicle with a mind of its own. All of these traits are critical. Few individuals possess them all. Yet all of them can be acquired through persistent practice.

Air combat is akin to four-dimensional chess where the practitioner must be a combination of psychology major, engineer, sharpshooter and martial arts fighter. Martial arts in particular have a strong similarity with air combat in that both disciplines makes use of terms such as energy (ch'i) and balance – in the air, you want to unbalance your enemy, catch him off guard, sap his energy and retain your own. You want to kill him with a single debilitating blow to the larynx, or cripple him by kicking his leg clean off. You will want to defend yourself by keeping out of his range and manoeuvre so that he cannot possibly hit you, reserving that opportunity for yourself. Or in the words of Morihei Ueshiba, founder of the martial arts discipline Aikido:

Move like a beam of light:
Fly like lightning,
Strike like thunder,
Whirl in circles around
A stable center.

And

Left and right,
Avoid all
Cuts and parries.
Seize your opponents' minds
And scatter them all!

Your physical and mental state is of fundamental importance when engaging in any sort of piloting. Fighting when tired or preoccupied is nigh impossible, and, what's more, not a whole lot of fun, which, at the end of the day, is what simulated air combat is all about. Having sortied several times almost daily for eight years I can safely say that the times when I felt stressed, angry, worn out or dejected, were the worst and would almost certainly lead to another swift death. Whereas the times when I felt centred, relaxed and pleasantly taut, alert and secure, usually ended in successful sorties. When in a good state my SA was supremely attuned, my moves deft and precise, my decisions instant and lethal, my actions decisive and full of purpose. When in a bad state, I might as well have stayed on the ground, for everything I did was sloppy, incomplete, half-baked, tired.

A good pilot, however, is not (merely) a guy who can scatter his enemies at will but one who is economical of force and resources, helpful and of good cheer, dependable and straightforward too. When flying as a team you will want to have a partner whom you can trust to act intelligently and predictably, promptly and safely. And you will want to fly in the same manner yourself, for your partner.

CHAPTER 11: ON GUNS AND GUNNERY

"I fly close to my enemy, aim well, and of course he falls down"
Herr Baron Manfred von Richthofen

With one notable exception, the 45-degree upwards firing "Schräge Musik" installation in Bf-110s and Ju-88 night fighters, pursuit planes employ fixed forward-firing machineguns and/or cannons to defeat the enemy. In order to score a kill, one must first fly these guns to a position where the guns are effective and then fire a concentrated volley that kills the pilot and/or permanently disables his plane. That much is clear. The bone of contention is how to get to that point with the least risk to yourself, and how to make that firing pass as devastating as possible.

It is convenient to discuss gunnery in terms of an aircraft's guns envelope – that small sector of the airspace in which the guns may deal out a disabling volume of fire. For the machinegun-equipped fighter, this envelope is small indeed: no more than 3-400 meters forward of the plane. For the cannon-equipped aircraft the envelope is somewhat longer, up to 6-800 meters, but the probability of a hit at such extreme ranges is not very great and generally requires a forbidding ammunition expenditure. The guns envelope decreases yet more against a target which refuses to play sitting duck: if the enemy manoeuvres, you will have to close to minimum range to increase the odds of a hit. The effective guns range is generally no more than 150-200 meters, and you will often find that closing to 100 meters or less is advisable.

When considering the guns envelope you must also factor in a few degrees around the boresight since the shooter may correct pitch and yaw while firing, and because gravity loads causes his fire to arc below his line of sight. Imagine the guns envelope as a cone with its highest lethality closest to the centre and with a reach of no more than 3-400 meters. As long as you stay out of this cone, nothing can hurt you. I state this truism deliberately because it isn't uncommon for pilots to be totally rattled by having a bunch of enemy on their six – if they're not in range, they're not a problem, so relax! Similarly, if you KNOW that you're turning outside the enemy guns envelope, there is no reason to be alarmed even if there is a horde of badguys chasing you.

Concentrating your fire is of the utmost importance. Machinegun-equipped fighters with one or more pair of guns mounted outboard of the propeller suffer from having their fire scatter in front of and beyond the convergence point – the point in space where fire from all guns are sighted to converge. This means that, if your convergence is set to 200 meters, your fire will be more or less spread out inside or outside of that range and thus the great majority of fire will miss the intended target despite it being fully inside the sight picture. Machinegun fighters scatter their fire in a "sheet" or "shotgun" pattern and thus it's highly advisable to fire only when the target is at or near convergence. Since that isn't always possible, particularly inside convergence, you'll find that you will have to "walk" your fire across the target or work to concentrate no more than one or two pairs of guns at the most vulnerable spot. It is also quite frequent, for beginner pilots especially, to become excited in the firing instance and further scatter their fire through hamfisting of the stick. If you're eager for the kill the higher the probability that you will hose your fire all over the sky, leading to even less hits. If you've experienced such excitement and its commensurate lack of success, you must learn to relax – lean back, control your breathing and loosen up your grip on the stick. Also know that anger, frustration and desperation impacts your gunnery negatively.

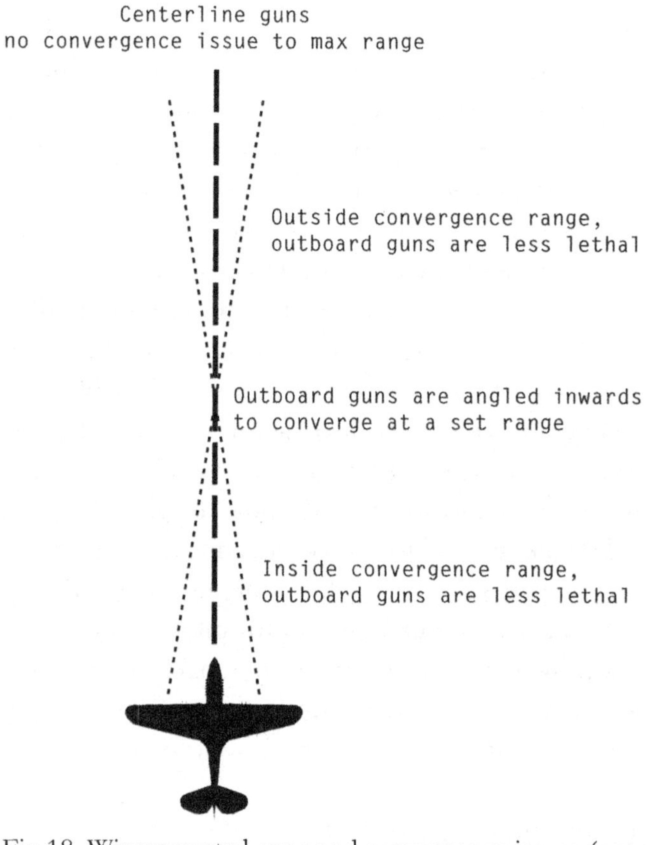

Fig 18. Wingmounted guns and convergence issues (not to scale).

To be a good shot, you must be calm and calculating, dispassionate yet concentrated. However, do not let your calm possess you to the point of holding fire in anticipation of the "golden" setup. If you do, you're highly likely to miss the "killingest" opportunity and may even end up tailgating the bandit. The thing to strive for is a detached, natural, state of mind that automatically seeks the best solution – then gunnery will become easy and effortless.

Learning how to shoot well is arguably one of the most challenging processes for the beginner pilot. Aside from constant practice and understanding of the forces involved, it takes a conscious effort to be calm when one is chasing a wildly bucking enemy or setting up a sweet bounce – adrenaline, sweating palms and a thumping heart contribute to mess up your gunnery. Be aware of your excitement! Lean back in the chair, take a deep controlled breath and loosen your grip on the stick.

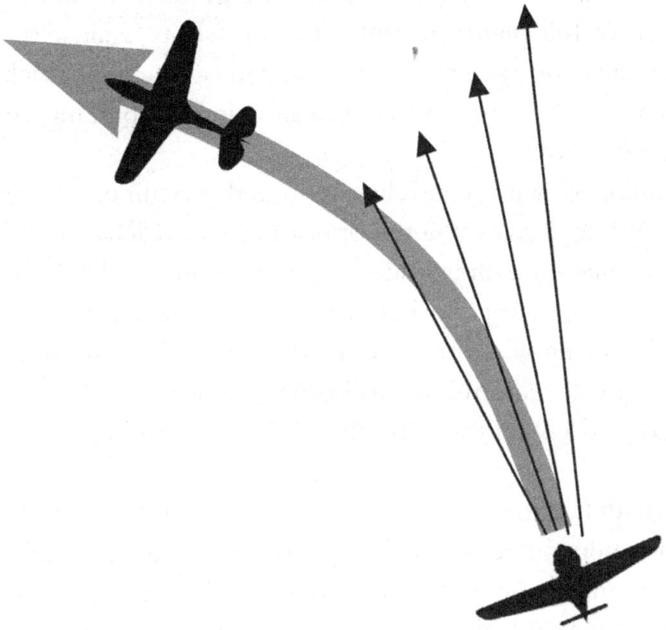

Fig 19. When pulling gravity loads (i.e. turning) your fire will arc behind the bandit even though you are pointing straight at him.

Another important thing to consider with regard to concentration of volume is to fire "unloaded" – i.e. to avoid pulling gravity loads when firing. Even the smallest back pressure on the stick will cause your fire to arc below the line of sight, further diminishing the lethal volume and most likely causing the burst to miss altogether. Thus, a "perfect" guns solution requires you to fly a straight line without the slightest elevator input up to the firing point and until you cease fire. This is the most crucial aspect of setting up a good shot: to fly the plane so as to avoid pulling G's before and during the firing sequence. That is a tall order in a close-in fight, and can be enough of a challenge for the energy fighter too. To this end it helps considerably to engage the enemy with lead pursuit, and to engage with a relatively small angle off tail lest the target disappear entirely below the bonnet. In the close-in fight you will likewise have to pull a good deal of lead (deflection), relax momentarily and fire, watch the enemy fly through your stream of fire, then haul back from the resulting lag pursuit situation in order to set up a new lead pursuit.

In a turning fight where you're not down to stall speeds yet but have the luxury of playing with your food, it is often comfortable to go to lag pursuit in order to create subsequent high angle off lead pursuit situations, rather than to "haul back and reel him in". Think of it as displacing the circles slightly. This is especially useful against slower targets, as it maintains and even improves your energy state. What you get then is a series of full or quarter-full planform unloaded snapshots against the bandit's engine and canopy. After each shot you're automatically in lag pursuit, which you convert to lead pursuit with a gravity assist (in the vertical going up, braking, or in the vertical going down, cutting the circle).

If your aircraft comes equipped with centreline-mounted machineguns or cannons, gunnery is much easier. While it gives you the opportunity to reach out and seriously damage the enemy at ranges exceeding the converging guns package, be wary still of using them at extreme ranges. It is far better to close to such a distance as to guarantee a lethal volume, this being when the enemy fills your sight entirely. If you close to minimum distance you will seldom have to bother about bullet drop or deflection shooting – all you need to do is fill the gunsight with enemy and light him up.

That said, being a master of deflection shooting gives you the opportunity to eke out effective hits from 3-400 meters range and high angle off. While getting really close is preferable, it isn't always an option due to insufficient closure. Likewise, snapshots at less than 50 meters range and high angle off are fleeting enough and carry with them the very real risk of collision. Snapshots are more often presented at 150-200 meters range, and at that range a fair measure of deflection is always required.

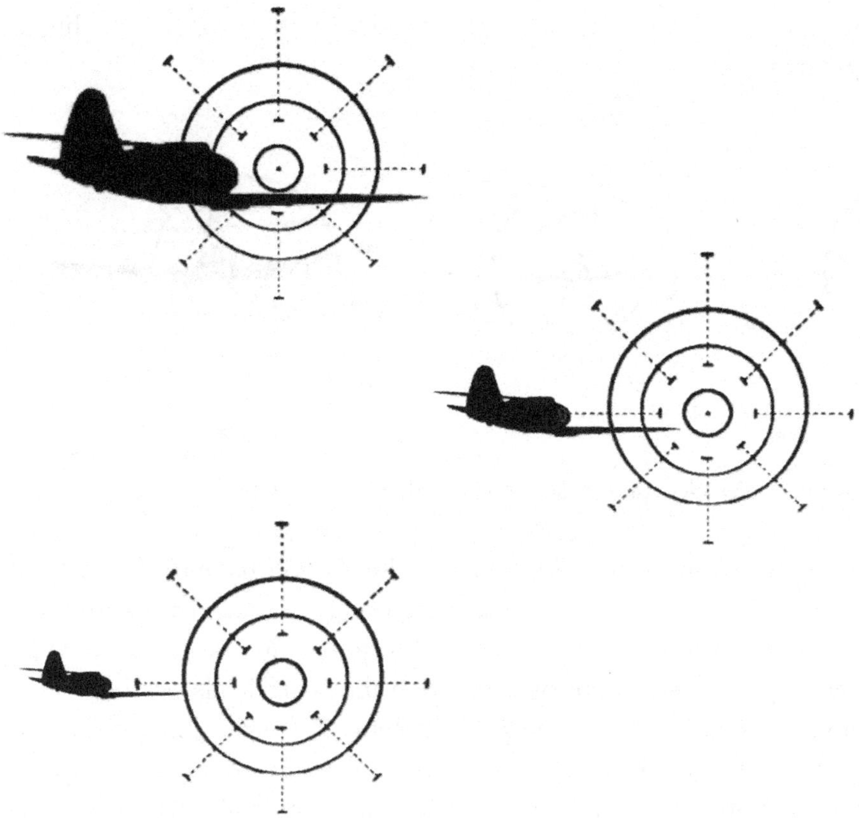

Fig 20. Range, and the rate at which the target crosses the line of sight, determines the amount of lead. The longer the distance, the more time you must allow for the bullets to reach the aiming point.

Deflection firing is difficult, no doubt about it. You will have to practice and practice again to find the ideal ranges and amounts of lead to draw to become proficient. When to fire and how much to lead depends on the range and rate of movement of the target across your bow – the longer the range and the faster the rate, the more lead you need to apply, up to a point where your fire is highly likely to be ineffective. The deflection solution is further aggravated if you have to contend with an oblique situation (both vertical and horizontal targeting element), and yet more so if you're faced with having to fire while inverted. Even at such short range as 200 meters, deflection shooting is beyond the capability of most pilots, so don't be disheartened if you miss a lot at first. There are no fast and easy rules of thumb, though, as a general guideline it can be said that if you engage at 150-200 meters and with a moderate closure and lateral rate, you need to draw about two-three full plane lengths at 90-45 degree deflection and about one to one half plane length at less deflection. If you're closer than that, aim just forward of the propeller to hit the engine and cockpit area.

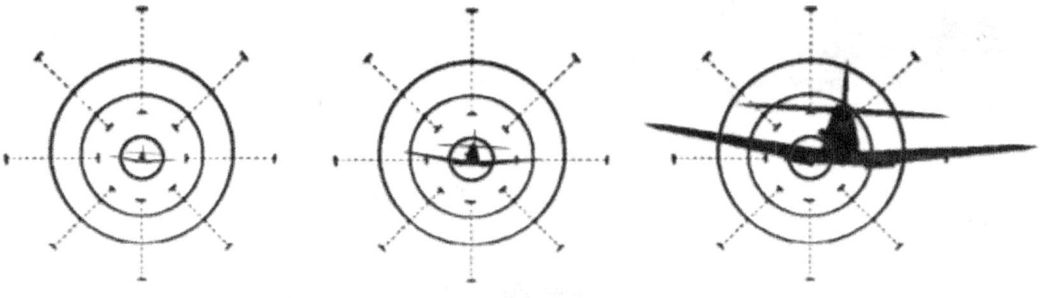

Fig 21. When you think you're close, go closer still and fill your gunsight with enemy.

For inverted gunnery (i.e. firing when flying upside down) you have to momentarily let go of your horizontal awareness and concentrate totally on the vectors involved. It can be disorienting at first, but when you get used to it, nothing beats the sensation of suspending spatial awareness. Think of it as firing in a barrel scissors fight - you get lots of near-inverted snapshots in that kind of fight. To make things easier you should endeavour to fly as straight as you can, preferably without any stick input at all. If you're pulling gravity loads while inverted and perhaps rolling a tad or even pushing on the rudder, your inverted shot will be nigh impossible to compute, particularly against a similarly manoeuvring bandit. So, keep the crate right, let go of spatial awareness and concentrate entirely on reading the bandit's flight path. Try to avoid long range shooting while inverted – the closer you go, the easier the shot computation.

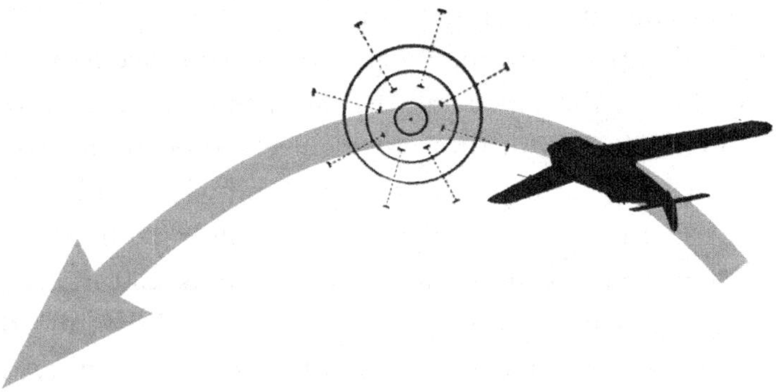

Fig 22. Coming in inverted and at a small bank (such as on top of an oblique circle) it takes a supreme effort to ease back stick pressure and concentrate on the bandit's vector for an unloaded deflection shot.

With regard to drawing lead in the bounce, the deflection computation is somewhat easier since your closure is generally extremely high, almost to the point of the target appearing to stand still in the air. Let's say that you're coming in high and fast on a relatively slow target. The bounce begins by flying unloaded (without pulling on the stick) towards a point some 6-700 meters in front of the target, in his projected flight path – this is the distance he will cover while you're closing in. Lead decreases naturally as range decreases. When the target's wingtips are fully outside the outer gunsight ring, open fire just shy of his nose and keep it up to rake the target canopy or put a devastating burst into his inner wing area from 100-150 m down to point blank.

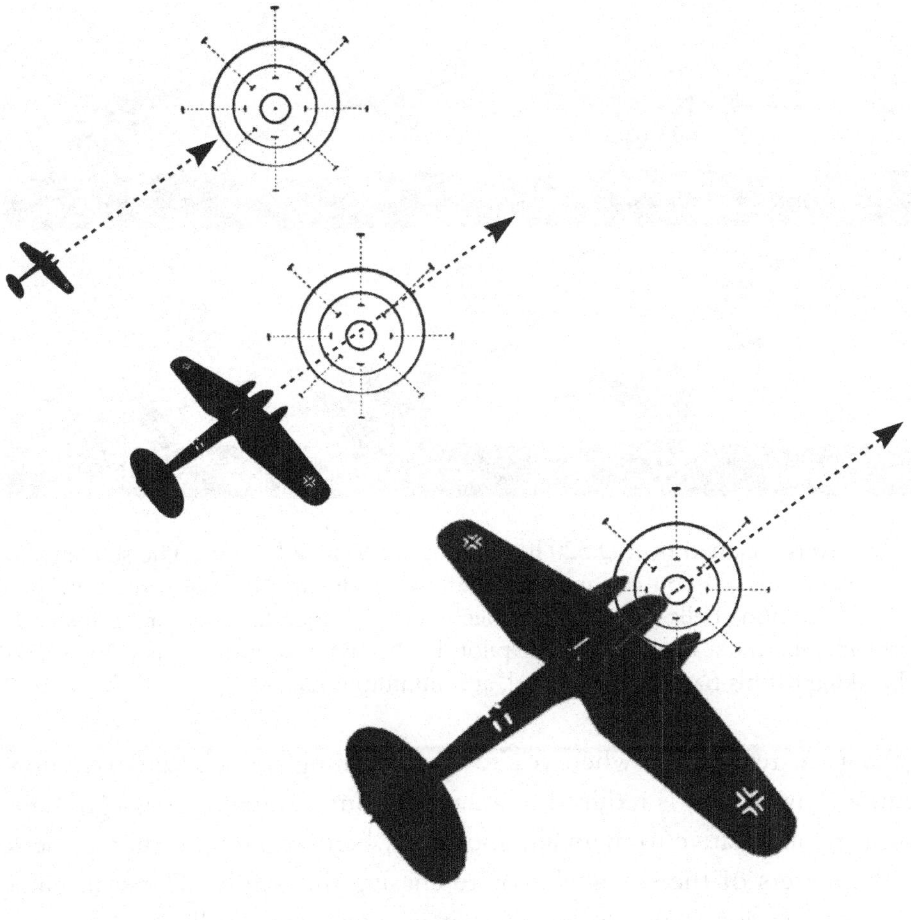

Fig 23. Take aim well in front and fly a straight course, allowing the target to fly into the gunsight. Fire at convergence and let the gunsight slide down along the fuselage and wingroot so as to avoid the collision.

Break off the pass by letting the target's speed carry him up through the gunsight, momentarily ducking below him as it were. Only if the target is really slow (as in a stall, climbing slowly or when close to the ground) should you break off by pulling up and possibly rolling a tad to get out of his plane of manoeuvre. Most pilots have a tendency to break obliquely UP, so ducking below is the good thing to do if your velocity vector is anyway in that direction. This of course is only advisable when you're coming in steeply - at other times it makes sense to break off high and roll to observe the effects of your fire, or zoom and bank to set up a chandelle plus a rope-a-dope, depending on the situation.

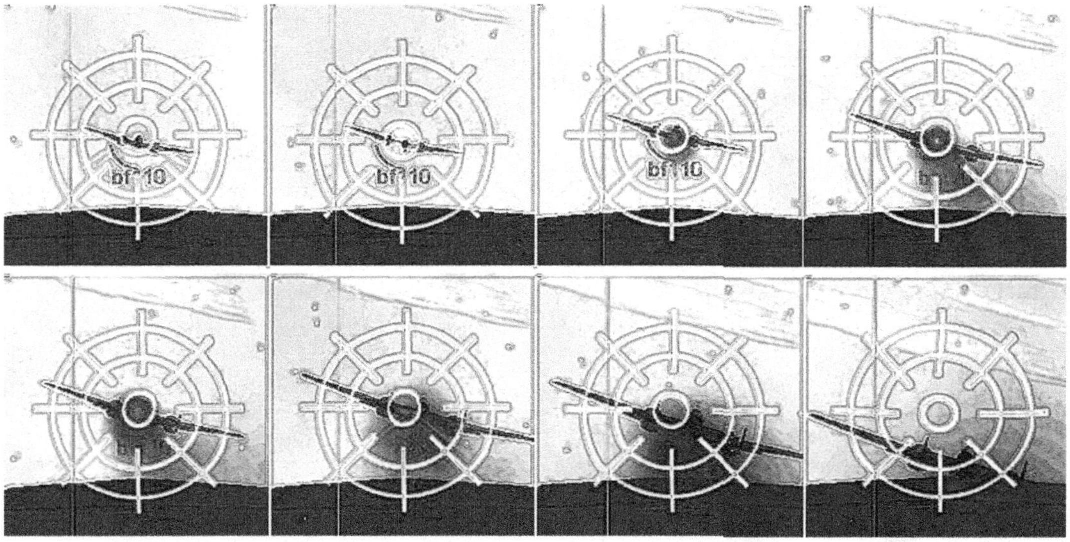

Fig 24. Here, a Bf-110 is creamed by a D.520 in a classic six o'clock bounce. The sequence is no longer than 2-3 seconds, so you can imagine the rate of closure. In this short timespan approximately 10-12 cannon shells and 200-250 machinegun rounds hits the target, instantly flaming the port wing and most likely killing the pilot. In the last two frames it is evident that the shooter is breaking off his pass in the vertical, at minimum range.

A couple of things to remember: when you're concentrating on your quarry, know that your situational awareness is reduced to a narrow cone defined by your gunsight view. It is absolutely imperative to maintain good SA when setting up a shot – check six, and scan all quarters of the sky while you're chasing the bandit. The scan takes no more than a split second and during that time your quarry will not get away. Immediately after a successful guns pass, break hard away in guns defence – for you are highly likely to have a bandit on your six.

WHERE TO AIM?

As a beginner and intermediate pilot you will most likely find yourself gunning at six o'clock to the enemy. The instinct to aim for the biggest part of the enemy plane is quite overpowering, but the centre fuselage isn't always the best part to aim for mainly on account of pilot armor. It is generally better to aim for a wingroot, or, if you're coming in with any offset at all, for the engine, the pilot or a fuel tank. When targetting a bomber it is generally best to aim for the cockpit first, secondly for the wingroots and thirdly for the engines. The bomber's centre fuselage is mostly empty air, so try and hurt something more vital first. Also remember that elevators and vertical stabilizers are critical flight surfaces – if nothing else is feasible, hurt him there.

And finally, another tip to improve your gunnery: don't use your zoomed in gunsight view (if any) but use your normal forward view when closing in to guns range. This will allow you to get closer, i.e. not fool you into thinking that you're closer than you actually are, it will help you avoid tailgating the bandit and makes it easier to gauge speeds and vectors in the guns solution. If you're addicted to the zoomed in view already it will take some unlearning, but once you free yourself of the zoomed view your kill rate is likely to skyrocket.

GUNS DEFENCE

The standard response against an enemy closing in on your six o'clock is… to panic. Without training, practice and a fair measure of mathematical logic, the neophyte pilot will lock up completely and remain a sitting duck when shot at. I'm here to tell you that the mere fact that someone is firing at you is no cause for alarm whatsoever. Panic is utterly uncalled for and you should learn to discern the difference between truly dangerous situations and those that present no risk whatsoever. Remember, though: if you can't see him, you can't fight him! If you can't see him, you know nothing!

Guns defence then, is dependent on several key aspects: you have to SEE your assailant in order to move profitably against him, you have to have ENERGY to manoeuvre against him and you have to mind your TIMING so that you move at the most auspicious moment. If you can manage that, defeating your enemy's attack will be a simple matter. Furthermore, you need to be CALM in your mind and STEADY on the hand, all the while working up adrenaline to boost your AGGRESSIVE SPIRIT. I should say that these key points really only matters, however, when you're beset by a formidable enemy enjoying one or several significant advantages – when defending against a lesser threat, the seasoned pilot manoeuvres reflexively and with little effort to maintain security.

The classic guns defence against a bandit coming in on your six with a head of steam is to make a radical break turn – a quick heading change – preferably with a vertical element thrown in (i.e. you turn and either dive or climb somewhat in order to force a more complicated gunnery calculation, one that he's unable to solve in time to score, on part of your assailant). Make your break turn just before he closes to guns range, and break towards the enemy. Similarly, if the bandit is coming in at you from, say, 4 or 5 o'clock, make your break turn towards the enemy, never away. The reflexive behaviour is anyway to break towards the enemy so you need not worry too much about that. Should you break away from the enemy, however, you're serving yourself on a hot platter: you'll present him with a simple stern shot. Key to guns defence is to present the enemy with a shot that is as difficult as possible, preferably one that he cannot manage at all. What happens after the first break is totally dependent on the situation. You might want to go into a barrel roll rather than to simply break fully away, you might want to reverse your break, you might want to disengage with a split-S – it's all up to you and the general situation.

The thing to remember is that guns defence is all about making the bandit overshoot (fly past) without getting a shot opportunity. Doing so is mainly a matter of introducing more angles to his guns solution than he can comfortably handle, i.e. getting out of his plane of manoeuvre[1] – of getting out of his performance envelope as it were. Depending on the speeds and distances involved, it may suffice to add a few angles to the situation, something that doesn't cost very much from an energy perspective. It is also important to consider the timing of the break – you don't want to advertise your new plane of manoeuvre too soon as this might give the bandit time to adjust his guns solution, nor do you want to break too late for obvious reasons. One other important consideration is the RATE with which you travel through his gunsight: you will want to move swiftly, i.e. at a good rate, so as to minimize your stay in his gunsight or so as to defeat his shot altogether. At other times you may need to manoeuvre radically to get yourselves out of his current and likely flight path – in such a situation it helps a lot to consider just how many G's the bandit may pull and where the limits of his angle of attack are at relative to your position. As soon as he's in front of you he has to set up anew, thus giving you freedom to manoeuvre against his continuing game plan – or even the opportunity to gun him down.

Other classic guns defence manoeuvres include the scissors, rolling scissors and defensive spiral (both up and down). Employment of these manoeuvres isn't limited to attacks delivered from your stern quarter – the scissor or lead turn works just as

[1] plane of manoeuvre in this regard refers to the comparison between your attitude and heading to that of the bandit. I.e. if you're banked to left and the bandit is banked to the right, you're in different plane of manoeuvres.

well when the enemy is to your side or coming in head on. We'll discuss these options more in the upcoming ACM section.

Remember though that guns defence is not a persistent state. Your objective is to defeat his shot, not to defend indefinitely. Every guns defence situation should be viewed as an opportunity to go on the attack – it only so happens that the enemy may be in a position to fire first. Defeat his pass and go on the offensive!

The trickiest guns defence situation by far is when the enemy is co-E and close in. Certainly it must be regarded as a complete failure on your part to end up in such a bad position from the outset, but them's the breaks from time to time. In that particular situation anything goes. Do anything and everything to get out of guns! Here it's a matter of proficiency versus proficiency, of reactions and of individual plane characteristics – if your roll rate is superior, use it to scissor or barrelroll the enemy! If you can pull more G's and AOA than the enemy or have a superior turn radius, use that to swing on his six! If neither is an option, try a skid! And if he still persists, try the "Hartmann Escape": stick forward in a corner and full opposite rudder. This will cause the bandit to red out from negative G's should he try to follow your manoeuvre.

A typical rookie mistake is to regard guns defence as something that is only intended to make the bandit's shot as messy as possible while screaming helplessly for help all the way home to mama. The result is often a pathetic flopping about and wishy-washy wing-waggling without any particular plan. Key to guns defence is to regard it as an offensive move, and this move must be executed deliberately and authoritatively. You can't afford to take your eyes off the bandit and simply panic – stay calm, watch his moves, feel his itchy trigger finger, reel him in and just as he thinks he's going to gun you down, make a single confident and aggressive move to reverse the situation and kill him before he gets a chance to collect his wits.

Guns defence takes many forms. In its most elegant form all you need to do is throw the bandit a few angles which he'll be unable to parry: increasing your turn rate or steepening a climb may be all that's required to stay outside his guns and performance envelope. The rookie opponent is likely to take the bait then and exceed his envelope, he's likely to fire his guns even though it's clear that he hasn't got a snowball's chance in hell of hitting. In both these instances you've won the fight already: you have gained all the intelligence you need to defeat him; you've forced him to blow his energy and/or position; you've forced him into a state of frustration and impatience from which he's unlikely to recover. Maintaining your calm and remaining in control of your aircraft throughout the most dangerous phases of guns defence is extremely important. You have to tell yourself that the guy behind you isn't dangerous at all, to the contrary, he's just another victim – he just doesn't know it yet!

Remember, always, that the enemy assailing you is actually at your mercy, however strong he may seem and however secure he may feel (the more secure he feels the more he fools himself!) as he comes boring in to gun you down. His idea that he is on top of the situation is complete at fault – all you have to do is disprove him. You are gauging his approach, setting him up and then elegantly sidestepping to capitalize on his incredulity and brashness. For as long as you see him coming, as long as you have the energy to manoeuvre and as long as you time your moves right, there is not a thing in the world that can hurt you.

PART III: ADVANCED MANOEUVRES

CHAPTER 12: SEPARATION AND TIMING

"I always thought to go around in circles, slower and slower, was a ridiculous thing. It's not the way to fight. The best tactic is to make a pass, then break off and come back."

Brigadier General Robin Olds, USAF

A lot of discussion between pilots and indeed the bulk of what has been said thus far in this book is centred on the fighter duel, the instance where two fighters of roughly equal capability meet in the sky and do combat. That situation, or at least the situation in it's pure and unadulterated form, is extremely rare: more often than not, the duel is interrupted by additional fighters, by ground fire, by loss of contact etcetera. Sometimes it's hard to single out an individual enemy for you are seldom alone in the sky and, in well-trafficked parts of the sky, you will have to compete with any number of friendlies to score the kill. Either way, the duel is the inferior way of fighting inasmuch it gives the enemy a chance to strike back. You should never allow that but reserve the right to strike solely for yourself. It is by far safer to pick off the unwary and the unseeing while they're enroute to what they consider to be the area of engagement, to set yourself up as bait for the unthinking and allow your wingman to swoop down and blast them away, or to use your wingman for that purpose. Only as a last recourse should you accept single combat. Regardless of where the bandit appears when the engagement starts you should be able to swing onto his rear with a single decisive manoeuvre and close the deal quickly so as to minimise degradation of your overall situational awareness. And as soon as you have spent SA to kill a guy it is highly advisable to break off at high speed and regain control of the airspace.

That said, it may sometimes feel like the only available fight is the melee fight, especially in the beginner's mind. While online air combat does tend to concentrate to certain localities where it normally degenerates into a melee, it is nevertheless so that all these fighters will have to come from somewhere, and in some instances, leave the furball and motor home. It is on the routes to and fro these concentrations that the intrepid fighter pilot can make his living, without having to risk all in a completely uncontrollable environment. If you can muster the necessary patience you will soon find that the opportunity to put bona fide tactics to use is abundant as long as you stay out of the furball, and that the only thing the furball has going for it is… itself. In short, if you want to learn, don't surround yourself with enemy.

Before we go into the mechanics of a fight, it is advisable to spend some time considering the topic of separation and timing. This is something most beginner pilots have no inkling about, which is readily seen by their propensity for always flying "pipper on the enemy". While closing with the enemy is certainly the objective, the bee line is seldom the most auspicious except of course when you start out behind your quarry. In all other aspects you need to factor in turn radius – the lateral or vertical distance required to haul around into the enemy stern quarter.

Consider the meeting engagement again where the opponents approach more or less head on. The vast majority of pilots will, without second thought, commit to a head-on firing pass – because this presents little difficulty in terms of tracking and keeping a visual on the bandit. The target is right there in front of our trigger-happy pilot and he doesn't have to manoeuvre much to put the pipper on his adversary. What he doesn't take into consideration is that the enemy is likewise presented with a simple shot, and the encounter is thus a 50-50 crapshoot at best. At worst, your gunnery sucks and you collide too, prematurely and unnecessarily ending your career. It is here, in the head-on encounter, that the experienced pilot uses his head and creates separation where none was at hand from the beginning.

As you can see in the following picture, the experienced pilot points his crate well to the side of the enemy, effectively setting up a scissors situation. It can also be regarded as the preliminary to a lead-turn merge, in which the less experienced pilot bores in for what he thinks is an easy shot. The experienced pilot is actually "giving away" angles for free here, only to brutally reverse the situation at the most auspicious time. It takes the less experienced pilot several fractions of a second to understand what is really happening to him, and when he does he will most likely react in the most predictable manner – by hauling back on the stick to keep his lift vector on the bandit – which naturally only serves to make him an all the more worthwhile target for our expert. In this little engagement wits defeats brawn, without the slightest risk for the experienced pilot, and with little effort. This ploy has two variants. In the first (fig 25), the expert pilot only makes a single guns defence break followed by a second reversal into the bandit's rear hemisphere. The bandit usually only picks up the first reversal, the one which ruins his shot at time 2. At time 3 the bandit is still wondering where the opponent went, while the latter is busy converting from lag to lead pursuit and an easy unloaded shot. By the time the bandit recovers he has but a few moments still to live incredulously.

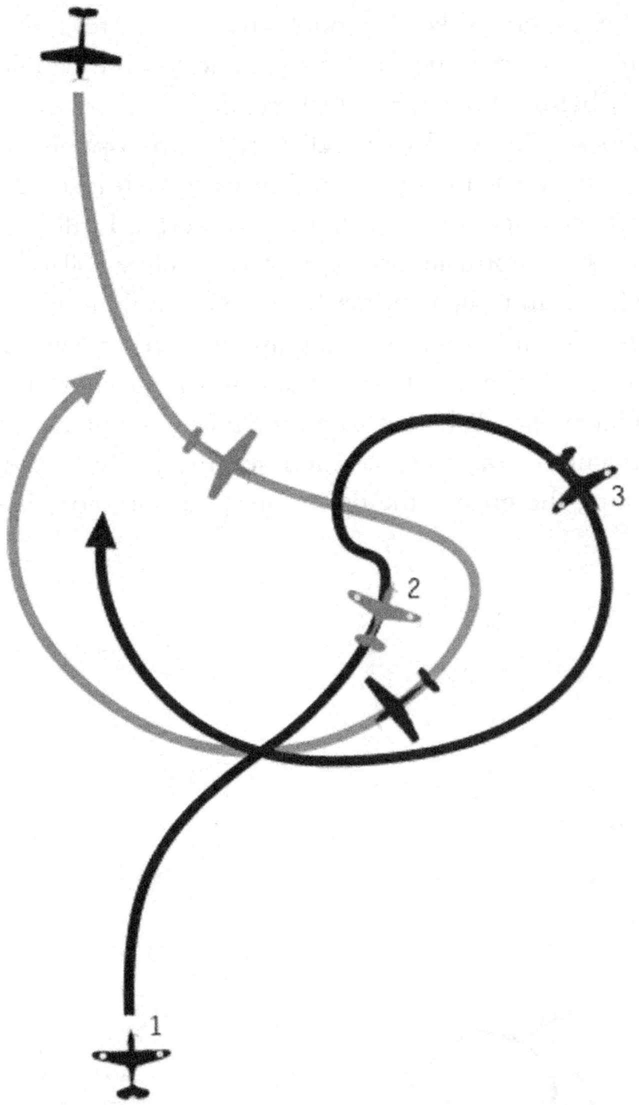

Fig 25. Setting up a predictable course of events. At time 1 the bottom fighter angles away to force the opponent into a predictable left bank. As the top fighter closes the distance he must continually adjust his lead and aggravate his bank. At time 2, just before the top fighter reaches firing distance, the bottom fighter, already in a left bank, breaks into and to the left of the opponent, forcing him into a predictable right turn. The top fighter must now reverse his bank from half left to fully right, causing him to lose precious moments. The top fighter is now convinced that his quarry is turning left and accepting a nose-to-nose fight. However, the bottom fighter immediately reverses his turn (preferably with a vertical element thrown in) and moves in for the kill. At time 3 the outcome is already decided: he is in lag pursuit at the apex of an oblique circle, and uses the gravity assist to convert to lead pursuit for the endgame.

In the second variant (fig 26), the expert makes a double reversal (across the projected enemy flight path) before the merge, the first into and across the bandit well out of range, and the last shortly before the merge, which rapidly takes him onto the less experienced pilot's six o'clock. This works so well because the opponent must always react to the new situation as it is presented, and in each such instance lose time and angles, and because reflexive action is so extremely powerful. I call this a "double fake" or a "frontal scissors". I normally use it if I have a clear roll rate advantage and against opponents who seem to be comfortable with brief high-aspect snapshots. In the latter case, the double fake serves to minimise guns computation time for the opponent: the less time I give him to set up and unload for his shot, the safer I am despite the perceived vulnerability. Whether to go for the single break or the double fake is generally dependent on the distances and speeds involved: the farther out and the slower the closure, the greater the opportunity to flummox the opponent with the double fake.

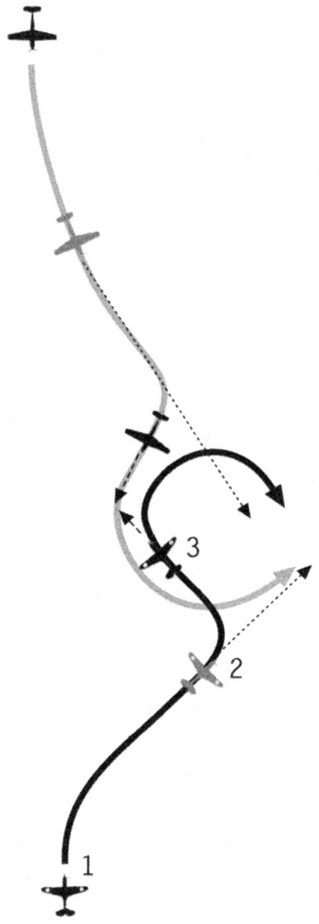

Fig 26. Using separation as a weapon (not to scale). At time 1 the bottom fighter eschews the head-on approach and angles out to force the bandit into a predictable turn. At time 2 the

bottom fighter swings across the bandit's path, forcing him to set up a new lead pursuit. At time 3 the bottom fighter starts another early lead turn which forces the enemy, still well out of guns range and a fraction too late, into a reflexive left turn. The bottom fighter, already well into his established turn, can easily convert to a full planform shot in the endgame.

Separation, and the diverging or converging vectors that ensure a certain separation, is the tool which you use to force the enemy this way or that, and the one thing that you need to keep in mind during engaged manoeuvring. When two fighters meet head on and commit to a turning fight, the fighter with the shortest turn radius and fastest turn rate will naturally dictate the engagement – because he will win the angles contest and climb onto his opponent's back. However, if the opponent doesn't commit to the maximum rate turning fight but blows past without pause, the turning fighter will have blown his hand by spending valuable energy for an unproductive position outside guns range. If only one fighter cares to engage, then he must ensure sufficient space to turn around so that he ends up in a productive position immediately following the reversal: this space is called separation, and the space requirement depends on your own speed and that of the target, i.e. the rate of closure. From this follows naturally that if you don't manage your separation, you forfeit opportunity.

Separation is also extremely important when discussing 2 vs. 1 manoeuvres and formation tactics. We'll explore it later on in the respective chapters, but for now consider the situation where you're fighting alone against two or more bandits: as long as you can keep them together to one side, or grouped up on your six o'clock, you can basically treat them as one. It's only when the bandits spread out (open up separation) and attack in sequence that it really becomes tricky – which is of course the objective for 2 vs. 1 manoeuvring.

TIMING IS EVERYTHING

Timing of various moves is arguably the difference between success and failure. The pilot's sense of timing; in the bounce, in the intercept; in the execution of reversals or any other situation which requires split-second accuracy, is, unfortunately, something which only comes with experience. One has to fail numerous times before getting a feel for the critical moment in a given situation. The subject of timing is highly relative to the situation: what are the respective energy states; what is the closure like; what are the respective turn radiuses at this speed and altitude; is the bandit inside or outside effective range etcetera. Fundamentally, the pilot must have a very good handle on everything pertaining to the situation in order to make the golden decision of when and how to turn. He must know his own ship inside out as well as the enemy's, he must accurately asses the energy picture, the enemy awareness and intent, and weigh these factors against his own proficiency and intent. In short, he must "feel" the situation.

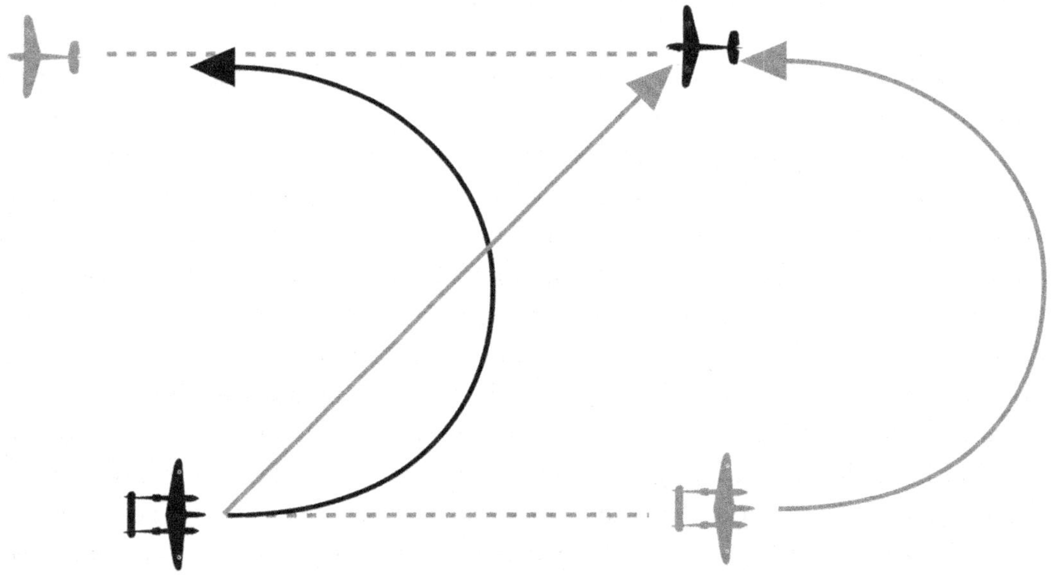

Fig 27. Plotting the turn, top view (not to scale).

Figure 27 shows the essential components of the timing calculation in a simple co-E situation: turn radius, distance and speed. Here, the attacking fighter starts his turn when the bandit is offset one full turn radius and at approximately 45 degrees angle off. This angle represents the time he needs to complete his turn at corner speed in order to end up at the bandit's six o'clock. If he delays his turn until the bandit is 90 degrees angle off, he will languish well outside guns range. On the other hand, had the bandit been a slow bomber, delaying the turn until 90 degrees angle off would probably be more productive. Clearly our bandit isn't likely to play along with the attacking fighter's game here most of the time, unless he's running naked for home, or if his SA is in the pits or if he's busy chasing someone (or even busy trying to catch up with his leader!).

The point of this is also to illustrate that separation is what matters. If the attacking fighter above had gone straight at his (unseeing) enemy, pipper on target, he would be saddled with a particularly tricky deflection shot of exceedingly short duration. In the case that the enemy is aware and reciprocates by going head on to the attacking fighter separation is nil, and most of the time nothing useful comes out of such a merge, except a fiery explosion and a lot of debris. In order to get behind your enemy you must consider your turn radius, and get into such a position so as to employ that turn radius to your benefit.

Another thing to remember is the vertical offset. If the bandit is several thousand feet above or below you will have to account for that fact when timing your attack, since you will get a gravity assist in the dive and trade speed for altitude in the zoom. When diving in you must consider what sort of attack you wish to execute: a high-side run with plenty of deflection; a high six; a level six; a low six, and so forth. You must consider what kind of speed and firing window you need, whether to make an immediate attack or a deferred one (i.e. a set-piece attack) etcetera. In any event the diving attack usually requires you to hold your wingover (the point at which you roll and pull onto your pursuit track) until the bandit has passed by your wingline, then you go in. Similarly, when dealing with a low-energy bandit above your level, you need to make a swift decision as to how to execute the attack, and initiate the attack run accordingly. Here your options are fewer, however, the challenge is none the easier: will you opt for a more or less straight low front aspect shot with a high six reversal behind the bandit (or a blowthrough); a level approach with a straight-up full planform shot; a level approach with an Immelman reversal into guns range, etcetera.

CHAPTER 13: ADVANCED COMBAT MANOEUVRES

"I didn't turn with the enemy pilots as a rule. I might make one turn - to see what the situation was - but not often. It was too risky."

General John C. Meyer, Vice-Chief of Staff, USAF

The ability to throw your crate around in the sky isn't what makes you a successful fighter pilot, nonetheless it's important to know how, and more importantly still, when, to perform one or the other aerobatic stunt. For even if your fighting style is the ultra-conservative, the high-powered boom&zoom variety, which should theoretically never put you in harm's way, you can bank on the enemy having other plans for you at some time or the other.

The moves explained here are seldom used singly, as is, but are usually used in combination as part of a grand scheme of manoeuvre in the particular engagement. That said, it behoves the aspiring pilot to make himself solidly familiar with each one so as to gain a feeling for his ability and the usefulness of the move.

Bear in mind that "dogfighting" is not the be all and end all of air combat. Verily, engaging in, or being forced to engage in, single combat (or if you're down on luck, one versus many) is a down and out failure. Not only is dogfighting THE hardest thing imaginable, for all its requirements of top-notch situational awareness and manoeuvring prowess, it exposes you to risks several orders of magnitude greater than that of a more conservative (but none the less demanding) engagement practice. Taking hits, or putting yourself in the situation where you might take a hit, is incontrovertible proof that you've failed to establish and capitalise on potential and real advantages, and ultimately, that your decision making and threat computing processes require urgent overhaul.

Going "fangs out, hair afire" into the fray may seem glorious, brave, aggressive and whatnot, but it isn't what combat is about. The truly consummate air to air gunfighter is he who never gives his enemy a chance, who kills without ever exposing himself to danger (other than that implicit by sheer presence in a combat zone), who never allows the enemy to get even remotely near a guns solution. As you become more proficient you can afford to ease up on your safety margins – to the point of engaging multiple enemy at close quarters without ever having to fear taking even a single hit. Until you reach such a level of proficiency however, do yourself a huge favour and engage only such enemy as you can safely handle, on your terms and on turf of your choosing. There is no dishonour in avoiding a bad fight, no shame in disengaging. Success is what counts, and that comes in many forms and shapes.

The manoeuvres described herein are not unique for the close quarters fight: they apply just as much to the energy fighter as they do to the angles fighter – and they can be seen both in the context of a highly localized and tight "personal" fight as well as in an engagement which covers tens of miles. A Chandelle, for instance, may be used to equal effect against a single bandit tooling around low over a frontline town as against a clutch of enemy chasing you for ten minutes and trying to get to grips at 30,000 ft. Keep these different scales in mind, and consider the amplitude, the scope and energies of various moves before deciding on what's appropriate for the situation.

13.1 SPLIT-S

The Split-S is used to reverse direction in a hurry with the aid of gravity. Roll inverted, and pull back. Recover in the opposite direction. Do it a couple of times and jot down your entry and exit altitudes at various speeds and throttle settings: this is your safety margin, and may just be the one ploy you need to make a less capable pursuer take an involuntary "dirt nap". Do know that the radius of your turn will differ considerably depending on your entry speed and throttle setting. In short, the faster you go the wider the turn. Also experiment with gravity loads – the more G's you pull the tighter your manoeuvre.

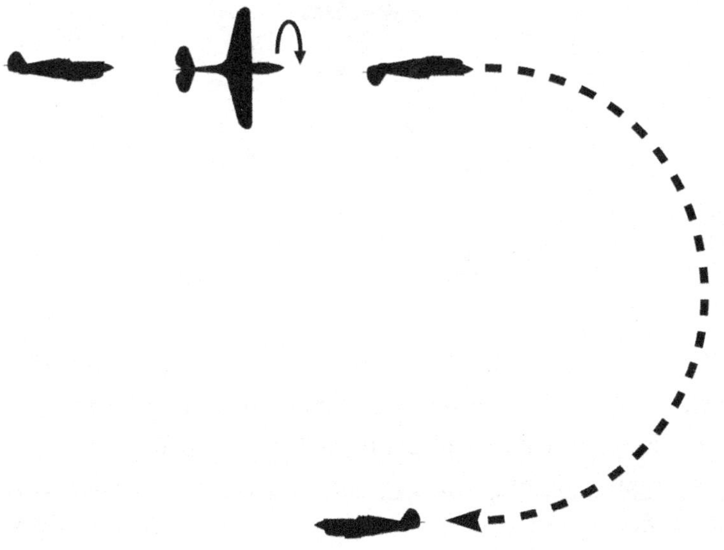

Fig 28. Split-S

Perform the Split-S at or below corner velocity since the gravity assist increases your speed considerably – at high airspeeds you will need more altitude to perform the same manoeuvre. A variant of the Split-S is to roll inverted, pull back to true vertical, then quarter roll and pull back to level. This will produce an exit heading 90 degrees left or right of your original heading. This is useful if a bandit on your tail can be seen to lose visual contact while you perform the manoeuvre. He might pick up your

Split-S, but fail to notice the additional heading change. Typically, the Split-S is used as a guns defence expedient, or as a preliminary to getting the hell out of Dodge. However, you should be extremely wary of using it as such if the enemy roll rate is superior, and you WILL be exposed to a full-planform shot if the enemy is deft enough to cut your circle, either by steepening a dive or following your roll. In other words, the Split-S alone is a rather predictable move and it's not likely to take you permanently out of trouble, to the contrary.

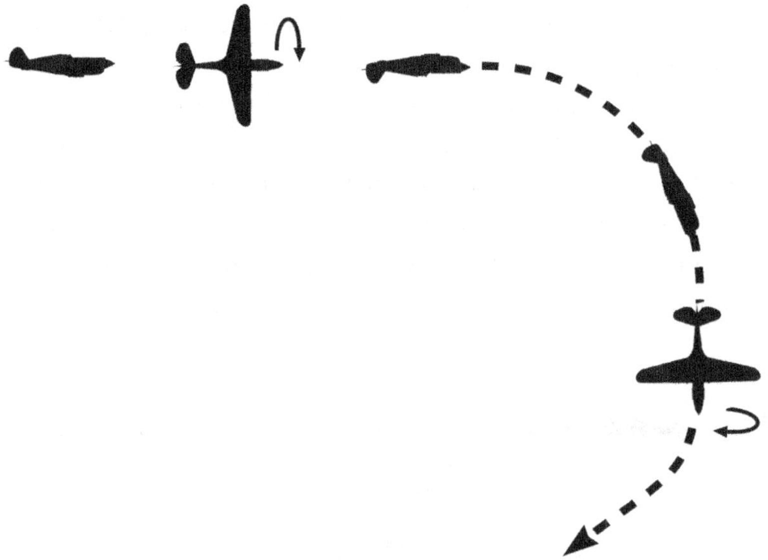

Fig 29. A half Split-S followed by a quarter-roll and perpendicular exit.

13.2 IMMELMAN

Named after legendary WWI ace Max Immelman, the Immelman is the opposite of the Split-S: it reverses your heading with a vertical displacement, but this time you fight gravity. Pull back as for a loop, but half-roll on top of the loop to exit in the opposite direction from your original flight path. If your airspeed is below corner speed at the beginning of the manoeuvre, you will have to make a short dive prior to zooming so as not to stall out on the way up. Like the Split-S this manoeuvre isn't much of a combat move per se but rather an expedient for reversing your heading somewhat faster than you would in a flat turn. The Immelman Turn can be a rather spectacular and rapid way of scoring a kill against an opponent approaching head on, or when you're approaching unseen in the bandit's low forward quarter, where his view is obstructed by the engine or wing: treat the situation as a vertical lead turn, though be advised that you need to be at maximum speed even against a decidedly plodding target and the vertical distance, ideally, no more than one sustained turn radius. If the vertical offset is too high, or your speed advantage not great enough, you risk ending up outside guns range and/or in a stalled-out condition, either of which is unproductive and usually downright dangerous.

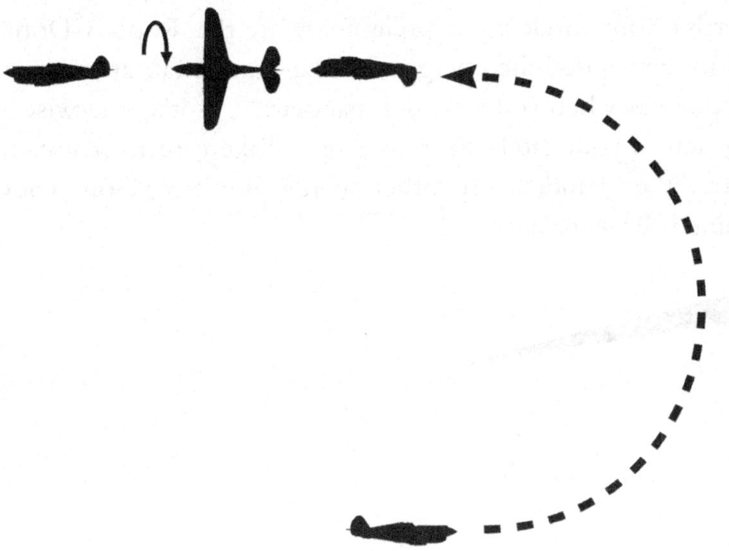

Fig 30. The Immelman Turn. Pull back and half-roll on top.

13.3 CUBAN-8

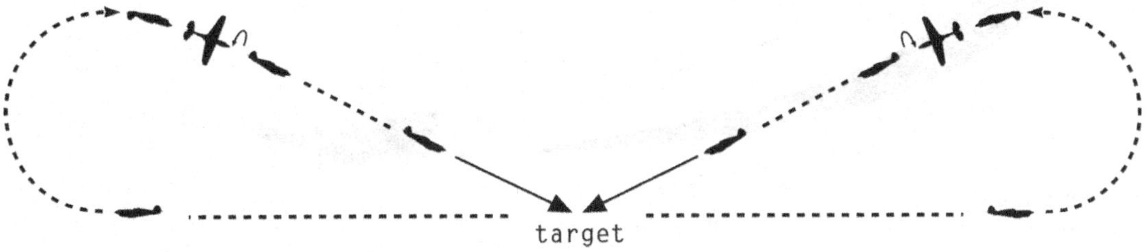

Fig 31. The Cuban-8.

The Cuban-8 is, as can be seen in the diagram, a figure 8 lying down, flown in the vertical plane. It's useful when strafing soft and squishy things, as it allows you to maintain a grip on headings while maximising your visual contact with the target. However, this comes at a price: you're a very predictable target for both enemy air and ground fire, so use it exclusively against targets that doesn't fight back. Make sure to extend a bit after your firing pass so as to avoid engaging the target in a series of loops. The Cuban-8 is a rather delightful manoeuvre as long as you're left undisturbed, and it also retains your energy and situational awareness very well compared to horizontal turns.

13.4 CHANDELLE

The Chandelle is a gentle climbing turn which reverses your heading with lateral (horizontal) and vertical separation from your original flightpath. It's a good way to induce a bandit to cut across your circle as a preliminary to the Rope-A-Dope discussed below. You need to have a definite energy advantage and a fair amount of separation to any would-be pursuer when you start the manoeuvre, as it's otherwise a rather simple matter to cut across your circle to guns range. Take care to maintain manoeuvre speed throughout the chandelle. In other words, don't max out your climb at the price of becoming near-stationary.

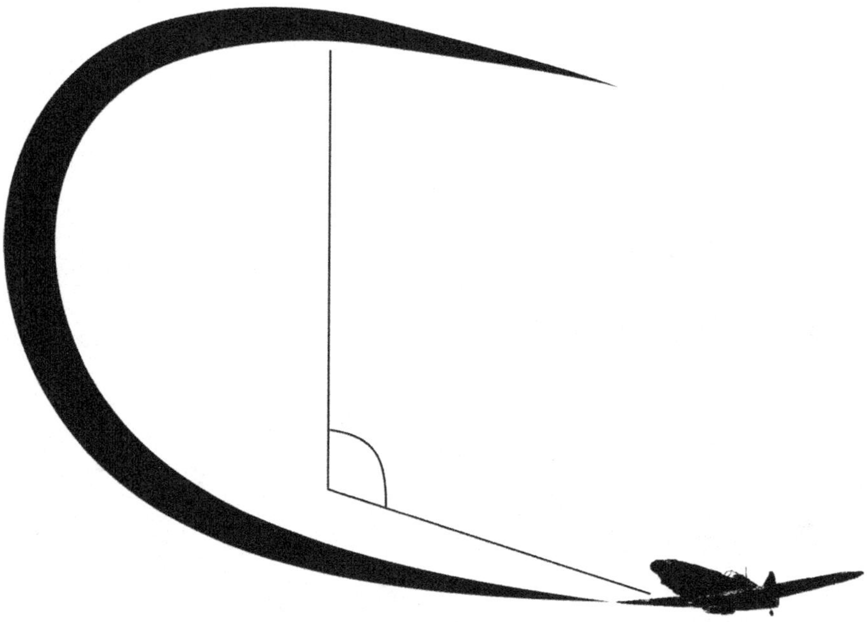

Fig 32. Chandelle

13.5 THE ROPE-A-DOPE

The Rope-A-Dope is a very handy manoeuvre when you have a large energy or performance advantage over a bandit engaging from behind and below your wingline. Against a much better performing plane, you can only adopt the Rope-A-Dope with a substantial energy overhead, and only for a limited time unless you kill him quickly or bleed his energy well between passes. In the graphic below, the grey fighter is decidedly lower and slower or otherwise suffering from a performance disadvantage while the black fighter is extending at a fair clip, e.g. from an aborted attack.

Fig 33. Use the Rope-a-Dope to drain the enemy of every ounce of energy.

At time 1, the grey fighter is in pursuit, well out of range. With sufficient energy in stow the black fighter starts an aggressive chandelle – the grey fighter follows as is his reflexive behaviour, trying to keep his speed up by putting his velocity vector slightly below the enemy (i.e. lag pursuit). At time 2, the grey fighter has used most of his energy to reduce the distance and is well into a maximum zoom to gain a tracking shot on the black fighter, in black's low quarter astern. At this point the grey fighter has gone from general pursuit to actively seeking a shot as he's closing in to guns range: he's adjusting to draw lead, momentarily forgetting that doing so will drain him of all remaining energy. Meanwhile the black fighter is maximising his spiral, getting every ounce out of his crate. The grey fighter is desperately trying to get his guns on him, and can almost but not quite make it. Here the grey fighter will usually open fire, either on hope or in dejection. As long as the White fighter retains enough forward speed and sticks to the turn he will have no fear, in fact, he should be laughing out loud.

Shortly afterwards the grey fighter mushes and stalls out. The black fighter spots the onset of his stall and is already swooping in for the killshot. At time 3, the grey fighter is recovering from his stall and have most likely lost visual on the black fighter, who is now in guns range of an amply predictable target.

Key to the rope-a-dope is to keep the enemy at or in front of the wingline during the final phase where he sheds every ounce of energy in the climb. What you do then is essentially a gently climbing circle around his velocity vector. Once you see the bandit mush or stall, or break off, you should already be on your way down. If you hold off for too long you give the bandit opportunity to recover manoeuvrability, so stick it to him before he gets a chance to compose himself. On the other hand, if the vertical displacement is insufficient for you to wing in for a stable guns solution, you might as well keep trucking (i.e. save your stashed energy and improve on it still more) and prepare to set up another rope-a-dope after he recovers.

A common error for the defending fighter in this scenario is to commit too soon to the attack. You mustn't dive in until you can affirm that the bandit has fallen for your ploy and is about to spend the very last of his energy. If you go in too soon the bandit will still have enough smash to point his guns at you, which will, if it doesn't kill you outright, cause you to prolong your dive past the merge and cost you dearly in terms of energy and position. For it is by no means certain that you have sufficient spunk to clean up and gun down the bandit on your first pounce. The rapidly developing guns situation calls for acute judgement and great skill as you will only have a few brief seconds to recover from the manoeuvre into the dive and frame the target in your gunsight, before opportunity is lost.

DEFENDING AGAINST THE ROPE-A-DOPE

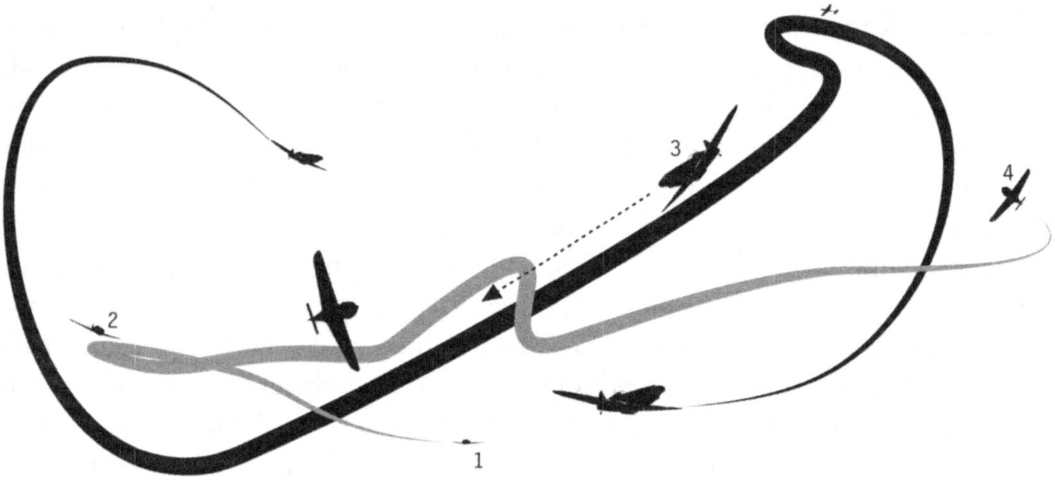

Fig 34. Defending against the Rope-A-Dope turns on not accepting his terms.

Defence against the rope-a-dope is pretty uncomplicated, but it does hinge on your ability to recognise the enemy intent early on, to restrain yourself from the reflexive response and of taking the appropriate steps. Figure 34 describes how the defending Black fighter refuses to play along with the bandit and instead forces him to fight on his terms despite starting the fight at an energy disadvantage.

At time 1 you're in pursuit of a extending bandit, at roughly co-altitude, who has a definite energy or performance advantage. However, instead of going right after him in the sucker chase described previously, you go into a wide lag position well out on his flank, in this example to his right. At time 2, the black fighter has increased his vertical advantage considerably, thus slowing his forward progress, in anticipation of roping in a dope. You however keep your distance and your airspeed, climbing only slightly, preparing to meet the onslaught. While the black fighter is at the apex of his chandelle and is beginning to realise that you didn't take the bait, you keep on climbing at a very moderate rate, effectively starting to level the energy disparity while the bandit dallies. You endeavour to keep the bandit well to your side, somewhere between your wingtip and your 45 degree forward view. You must at all cost resist the urge to point your nose at him.

Prior to time 3, the black fighter has decided to attack since you appear to be extremely benign. You see him commit to the attack (he briefly shows his planform as he rolls into you) and comes a-boring right at you. You reciprocate by turning into him, although the timing and amplitude of your move is highly dependent on the distance and closing rate of the bandit. The lateral separation you've created is now forcing the black fighter to spend some of his smash in closing the distance. This evens out the previous energy disparity yet further and transforms the engagement from a rope-a-dope to a frontal scissors engagement. Again, range and rate of closure, as well as altitude offset, determines whether you may employ a double fake or merely a single break. In figure 34, the defending fighter makes but a single guns defense move at time 3, probably a barrel roll around the black fighter's guns envelope.

Following the missed shot the attacking fighter will either a) extend or b) pull hard towards you or c) zoom to immediately reengage. Your options here are to a) reverse your turn again and get on his six if the energy margin is small, or b) keep trucking to reinstate the horizontal separation. Don't fall for the sucker trap of trying to eke out a shot as the black fighter zooms – that will most likely set you up for the rope-a-dope endgame as discussed above – it is far better to use your speed to rebuild the lateral separation. However, if the bandit does turn hard after you, dumping his smash, the wise thing to do is to reverse toward him (unless you did so already) and enter a rolling scissors fight.

In figure 34 the black fighter recovers after his missed shot, though in the process burning yet more energy by holding his dive for too long and by pulling too many G's in the subsequent transition, and goes into a wide oblique turn to right, while the defending grey fighter recovers efficiently from his break and hauls around for the continuing match at time 4. The black fighter retains sufficient smash to keep the advantage, but the black fighter is now much better poised to take the fight to neutral ground provided he doesn't succumb to the temptation of a head-on shot.

Key to dealing with the rope-a-dopeing bandit is clearly to not play his game. By adding horizontal separation you demonstrate that you're not falling for his ploy and force him to play according to your rules. Continue calling his bluffs and continue to drive home the point that you're in command of the proceedings – not he!

13.6 DEFENSIVE SPIRAL

The defensive spiral is used to force an overshoot of a bandit on your tail while in a dive, allowing you to either kill him then and there, or to disengage from the fight. The object of the spiral is to slow down your descent as much as possible – chop throttle, increase drag by increasing your Angle of Attack and deploy flaps momentarily. Unless the bandit is very good he will not notice your braking action in time and simply drop below you. The defensive spiral works best if you allow the enemy to get close before slowing down and "dirtying up" (i.e. increasing drag) – the less time he has to determine your action, the better. Think of it as a rolling scissors in the vertical, where you let gravity do the greater part of the job. This manoeuvre requires a lot of altitude if you enter it at top speed – don't try it below 3 km (9000 feet). Maintain visual contact with the bandit throughout the manoeuvre and turn so as to position the enemy off your wing rather than off your tail. This will induce him to try and cut across and thus overshoot all the sooner. The more he slides forward in your view, the better the spiral is working.

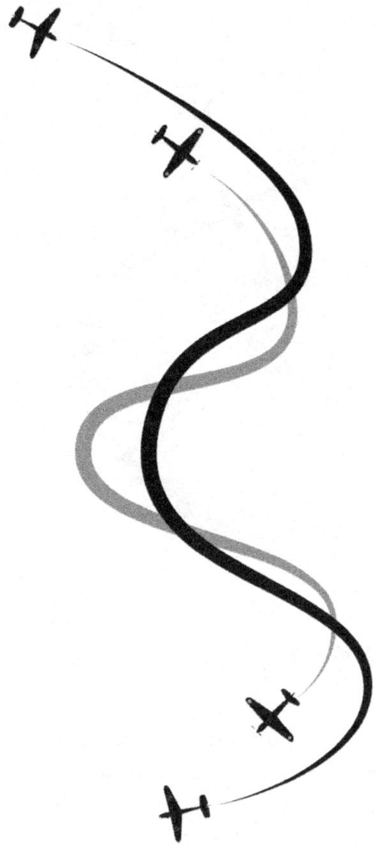

Fig 35. Defensive Spiral, side view.

The defensive spiral or Corkscrew works just as well going up, particularly against a bandit who comes in with a lot of energy or has a higher wingloading than you do. You need to be at corner speed or better to initiate the corkscrew however – in all other cases, a simple scissor or break reversal is likely to work better. The Corkscrew is essentially an oblique spiral where you allow the enemy to cut across for a guns solution. However, since you're both climbing and turning at max rate, in a spiral that is not truly vertical but leaning to the side somewhat, you are constantly feeding him a more difficult guns equation, up to the point where he overshoots. At that juncture you should already be turning onto his high six o'clock. The trick here is to position the enemy off your wing while spiralling up, i.e. if you keep your wingtip pointed at the enemy while you zoom you will automagically manoeuvre around him until such time as you see him stall out or flash past behind you. If the Corkscrew doesn't put you onto his back after a single 360 climbing turn, then you're doing something wrong, e.g. committing too soon or too late to the spiral.

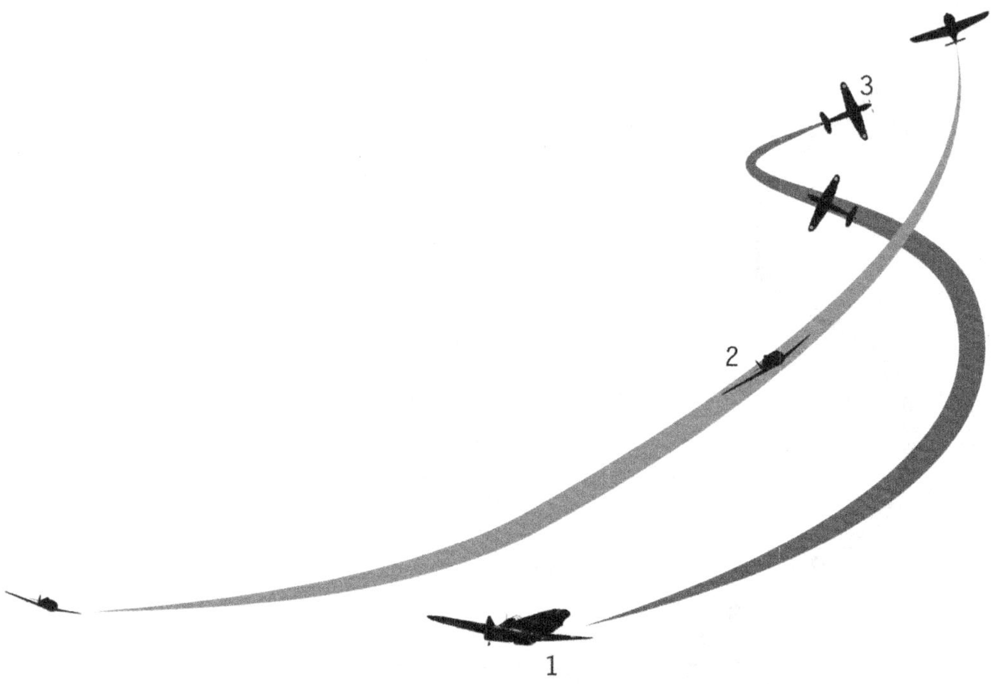

Fig 36. The Corkscrew involves a climbing and rolling barrel manoeuvre around the attacking fighter's performance envelope. At time 1, still a few seconds from getting into the attacker's guns envelope, the defender goes into an aggressive oblique spiral to left that produces a 90 degree deflection shot at a good lateral rate at time 2, forcing the bandit to pull hard to gain an impossible lead pursuit solution and thence to release to avoid the black-out. At time the defender has completed a 360 degree turn and has a good snapshot opportunity. The more the attacker pulls at time 2, the better the guns window the defender has at time 3.

It takes a certain amount of practice to get the timing and the amplitude right in the Corkscrew. If you start it too late you will not generate enough rate out of the enemy gunsight or not enough rate to get out of his performance envelope. If you initiate it too roughly you may black yourself out and/or suffer an accelerated stall, with all the ugly repercussions. If you make the corkscrew too gently you will remain too predictable and allow the enemy a pretty simple shot. So, with all these caveats, it should be clear that you need to operate the Corkscrew at corner speed and at the most auspicious moment, with a good balance between the vertical and the horizontal rate. If you manage everything right you will force the enemy to constant and fruitless adjustment while reserving the right to strike for yourself.

In my experience the Corkscrew works best against an opponent whom you have already established as being of the turning kind, since you're committing to an essentially static fight (i.e. one that is concluded in a comparably small area). If your opponent is of the extending variety, employ the Rolling Scissor instead as this keeps up your momentum. You will also need to be in a crate that turns reasonably well compared to your opponent, or at the very least, not worse than the opposition. If you tend to stall easily either in the instantaneous turn or in the sustained spiral, consider some other manoeuvre as your bread-and-butter defensive gambit. I should also mention that trim is of considerable importance in the Corkscrew, since you normally enter the move from maximum level speed. The manoeuvre will burn energy at a fair rate, pushing you down to corner speed and possibly even lower: if you remain trimmed for max level speed while operating at near stall speed, you will not gain as many angles as you might have, had you trimmed accordingly. Therefore, as you slow down in the oblique spiral that is the Corkscrew, trim tail heavy.

13.7 HAMMERHEAD (AKA MILITARY WING-OVER)

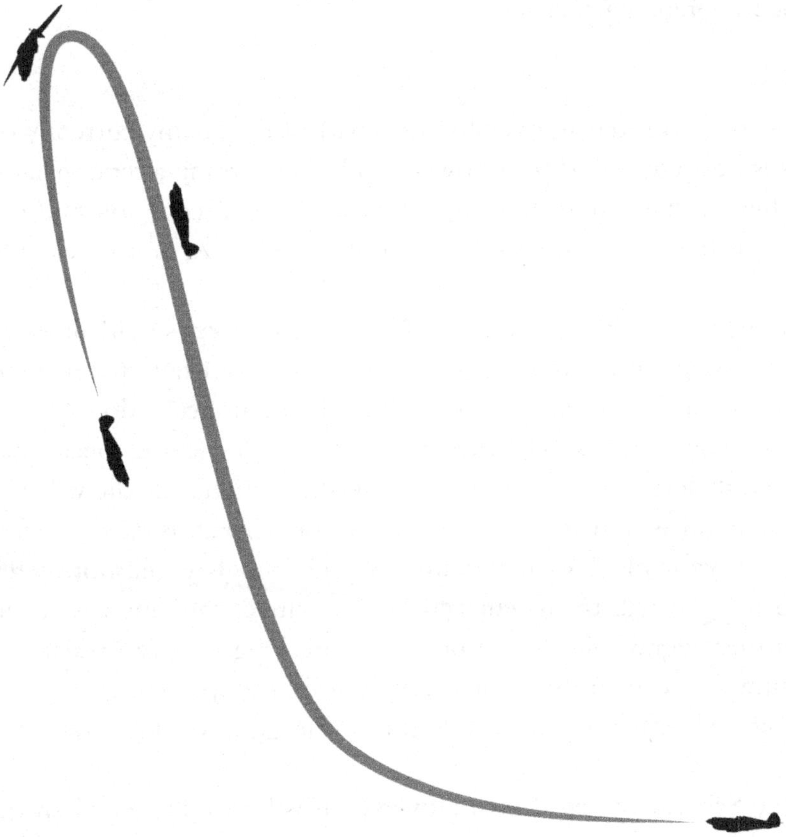

Fig 37. Hammerhead Turn.

The Hammerhead is useful when you have a lot of energy over your opponent or when making strafing passes at an acute angle. Pull up gently to near vertical and make the most of your energy. Look out to your side to determine your angle during the zoom. As you approach the power-on stall, tramp down on the rudder in your propeller torque direction (to give you a minor yaw assist) until you face Mother Earth. You may need to correct your flight path somewhat with rudder and a smidgeon of elevators, as the tendency is to exit the Hammerhead in a steeper angle than you entered it. The Hammerhead is not a bread-and-butter move that you can employ every so often, particularly against more or less equally performing enemy, as it is dependent on you having a significant energy advantage on the order of 2:1 – something which you may have against a bomber or when coming down on a slowly climbing enemy from several thousand feet of altitude. Against the latter however, it is probably more constructive to use your overhead of energy to sweep around rather than to set up the very steep high angle off shot produced by the Hammerhead. Furthermore, the Hammerhead is rather a conspicuous manoeuvre that takes some time to perform, which makes you a predictable target for anyone with a stash of energy and an easy foe to defend against.

13.8 FLAT SCISSORS

The object of the Scissors is to force an overshoot on part of the enemy currently on your six. The Scissors is best employed by a slower and better turning aircraft against a faster fighter with high wingloading, but any aircraft with a decent roll-rate can scissor. The Flat Scissor is basically a series of reciprocating reversals, or a series of S-turns if you wish.

The important thing to remember is that YOU are in the driver's seat, although it may seem contrary to perception. After all, the badguy is on your tail, therefore he should command the situation, right? Nope. All the enemy can do, beyond disengaging, is react to your moves. He can of course let loose the occasional snapshot, but it is nevertheless you who command the fight. The bandit will feel compelled to match your turns and in so doing he will be a fraction late in every reversal, unable to stay in your plane of manoeuvre (i.e. his wingline will not match yours). In addition to this, he will try to cut across your circle, thus increasing his forward speed and further aggravating his troubles. Before long (2-3 reversals) you will be essentially neutral, nose to nose or at the very least canopy to canopy – at which time the enemy will either overshoot before your guns or fall prey in a stationary turn fight.

You can enter the Scissors at any speed. However, it is best initiated when the enemy is already in or very close to guns range. If you start it too soon, all you will do is to present a comfortable target as the enemy closes the distance, cutting across your feeble and predictable reversals. It is also highly auspicious to make your reversals inverted, that is, instead of sticking to turning nose-over-the-horizon, you half-roll through the current bank to inverted and pull to reverse somewhat nose low as you complete the half-roll. This may cause you to lose sight of the bandit momentarily as you swing him through your low cold six, but it will also make your reversals more rapid.

Why does the Scissor work so well? It's because simple human reaction time works in your favour. You are the one who leads the enemy along – the pursuer must contend with the brief span between seeing, deciding and acting on a given input – and he is thus always a fraction late in matching your manoeuvre. The worse reaction he has, the better your scissors works, and if his roll rate is inferior to yours the scissors will work even faster.

Key to making the Scissors work is of course timing, but first you need to keep your eyes on the bandit! Your eyes should never stray from him, and since he'll be behind you and in your high rear a good deal of the time, you need to be adept at "flying backwards". Once you see his wingline getting out of synch relative to yours, you know the scissors is working – keep it up and aggravate the plane-of-manoeuvre disparity until you're canopy to canopy with him. You'll also see him creep farther forward in your upview during the reversals, indicative of your forward progress being slower than his. Keep it up, and you'll soon find yourself behind his wingline – the table's reversed! Never forget that YOU are in command of the fight. You're not defending here – you're attacking!

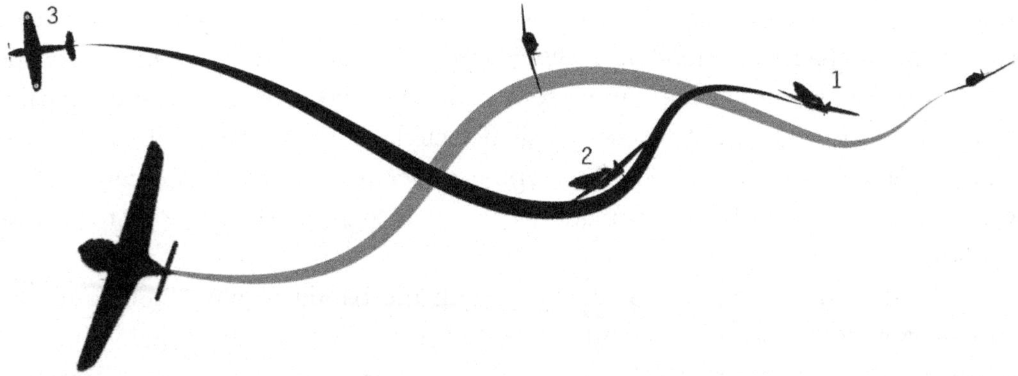

Fig 38. Flat Scissors (combat view). At time 1 the black fighter rolls left to bank his wingline opposite to the enemy on his six. At time 2 the grey fighter has reciprocated and also rolled to left while the black fighter is already well into a right-hand bank, pulling across the attacker's front. At this time the attacker has moved from the defender's dead six o'clock to his high five o'clock view – a definite improvement! – and the respective winglines are now banked in roughly opposite planes of manoeuvre. At time 3 the defending fighter has rolled to a left-hand turn again and come essentially neutral, canopy to canopy, while the attacker is still struggling to haul around. From here on the defender can easily convert to a decent angle-off shot as the scissors snaps shut in the subsequent instant.

A typical rookie mistake in the scissors is to make the reversals too gently, too mechanically and without observing the enemy. Each reversal is in effect a guns defense reversal and should be treated as such, not as a "mess-up-his-shot jink". It is also highly auspicious to start your next reversal at the precise time when you see the enemy labouring to follow you previous reversal – at that point his inertia and the inescapable reaction interval will make him lose that tiny fraction of an instant so crucial for success in the Scissors. While you're already halfway into your new direction, he will still be wallowing in the old one – and as he tries to reciprocate, you reverse again! Aggravate this wingline disparity and never ever take your eyes off the bandit in the scissors!

ROLLING SCISSORS

The Scissors can be either Flat (i.e. you reverse in the horizontal) or Rolling in the vertical. If you have the requisite airspeed to work in the vertical, a series of short zooms and dives in a barrel rolling manoeuvre allows you to slow down your forward progress more effectively and put yourself behind the enemy for a nice and safe guns situation. This is most taxing on your SA though – and the enemy's – so be sure to know that you can handle it.

It is quite common for the uninitiated to think of the Rolling Scissor in terms of rolling the crate as rapidly as possible, and that the better rolling aircraft is by default more suited to the Rolling Scissors. Not so. The Rolling Scissor is in fact a rather gentle manoeuvre that is more dependent on separation and closure management than roll rate. The Rolling Scissor is best initiated when the enemy is laterally offset, i.e. not already up close and firing on your dead six. One typical such situation is the plain vanilla nose-to-tail flat circle fight. When the bandit is clearly winning angles on you and slowly creeping closer for the finale you can, instead of waiting for the inevitable confirmation of your turning ineptitude, opt for a Rolling Scissor.

Thus turning, let's say to the right, with the bandit in your high four o'clock and closing, you nose down slightly to build up speed, then swing high while rolling to inverted, and continue rolling gently while pulling, now bottoming out sunny side up then continuing the roll while pulling again for a fresh rolling zoom. You have now left the static circle and are inscribing a corkscrewing motion, lying down, barrel rolling as it were. The bandit is of course trailing and trying to get a clear guns solution, thereby inadvertently cutting across your rolling circle: this is your opportunity. Open up the funnel as much as you can, still rolling and pulling only so much as to keep the wide barrel going, without losing so much steam as to become entirely predictable. You entice the bandit to dance with you. You will notice that he is creeping forward in your up-view, to the point of coming neutral, canopy to canopy. At this point you must restrain the urge to pull immediately into the bandit but instead continue the wide barrel roll. If the bandit pulls hard into you, evade the upcoming collision and scoot by him as close as you can. By now you're neutral and going almost in opposite directions – time to decide whether to continue the rolling, turning fight or to blow past and extend away. In either case, you have cheated grim death and gotten a new lease on life. Make hay!

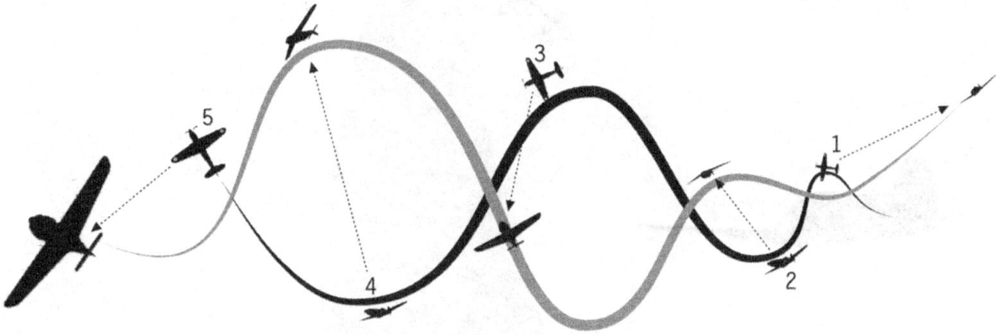

Fig 39. Rolling Scissors. At time 1 the defending fighter (black path) is in an energy-conserving barrel roll, corkscrewing around the attacking fighter's manoeuvre envelope (grey path). At time 2 the overshooting bandit has rolled to inverted to keep the black fighter in view, who has also rolled upright and is pulling into the attacker. At time 3 the black fighter has slowed down considerably compared to the attacker, who is endeavouring to keep his lift vector on the defender. At time 4 the fighters are again opposite, essentially neutral (canopy to canopy). At time 5 the defender has slowed down sufficiently to place him behind the attacker's wingline in his high 12 o'clock and can now easily pull in behind him.

You can also start the Scissor well out of range, and then you will hardly need more than two reversals to get on his six. The key here, assuming that you're being chased from the scene by a bandit well outside guns range, is to make a clearly marked turn to decrease the angle off tail (from dead six to say, 45 degrees off). This will induce the bandit to cut across your projected path, and set him up for an aggressive reversal once he draws almost into guns range. This second reversal will bring you on his six, the timing of which can be quite close to the first reversal. In figure 40 (not to scale), the grey fighter makes a radical guns defence reversal at time 1, and another at time 2, by way continuing the roll to inverted and pulling towards the black fighter – when the black fighter is still expecting him to be travelling in the heading indicated by his reversal at time 1.

Fig 40. Instead of two flat reversals you can instead execute a single rolling manoeuvre with a certain vertical amplitude over the bandit's flight path. The key here is to know just how much time you need to haul up in a guns-defensive barrel roll, and how much AOA you can pull at your present speed without blacking out or paying an exorbitant energy price for the manoeuvre.

As your SA, your confidence and your plane-handling proficiency improves, you will gain a fine feeling for the timing and amplitude of the breaks in the scissors. In the sucker chase shown in figure 38, you will find that instead of executing two standard oblique breaks, one after the other as you verify that the enemy behaves as expected, you might as well merge those breaks into one – into a rolling scissors move, or a barrel roll attack if you wish. This will preserve your energy and bring you much closer to a comfortable guns situation on the enemy – on the other hand, it does require top notch SA and excellent stick and rudder coordination. What you do, essentially, is to roll up and around his flight path, keeping him in view through you top canopy and letting his momentum carry him through, down to your forward quarter. At that point you're on a parallel heading but rolled inverted so it's a simple matter to finish the rolling manoeuvre and slide in on his six. The vertical offset is important here – make it large enough to give you a good guns defence at the beginning of the manoeuvre but small enough to retain your energy for the endgame.

How do you defend against a scissors? Or rather, how do you, as the attacker, retain your advantage? That too is quite uncomplicated. If the bandit is decidedly benign in his moves, you might want to cut across and gun him down in a single pass, ready to use the vertical to stay behind his wingline. If he seems to be an above-average stick, abstain from even trying to follow his scissoring and make a short rolling zoom well behind his wingline instead (to keep him in view) and wait for him to either tire of the flat scissors or to make a move against you in the vertical. From there on, you're likely to enter a classic rolling scissors fight. In such a situation the only advice is to keep you energy and stay behind his wingline until he makes a significant error for you to capitalize on.

Finally a small tip that may help you in the rolling scissor: when rolling with the enemy, try to fly so that you have visual on him in your forward-corner-up view rather than in your corner-back or straight-up view. This is important, as in the latter case you will subconsciously "contract the barrel" and remain essentially neutral down to collision range whereas if you place him in your forward-corner-up view you will subconsciously manoeuvre to slide around, or corkscrew, his flight path until you can haul in and saddle up.

I go into full aggressive mode. First, a couple of tight barrel rolls. Then, as the FW comes closer, sustained scissors with a measure of vertical thrown in. It seems to catch him unawares and on the second break I KNOW I have him. He gets totally out of sync on the fourth break and instead of extending, he turns in the horizontal. Wheeee! I see him struggle against the [stall] horn in my high twelve, and close in leisurely for the kill. I get a full two-three second guns window and make damn sure my deflection is correct before firing HARD in his high six, slam-dunking his engine, cockpit and port wing. He goes down, wingroot afire, and crash.

WarBirds, January 31 1999

13.9 LEAD TURN

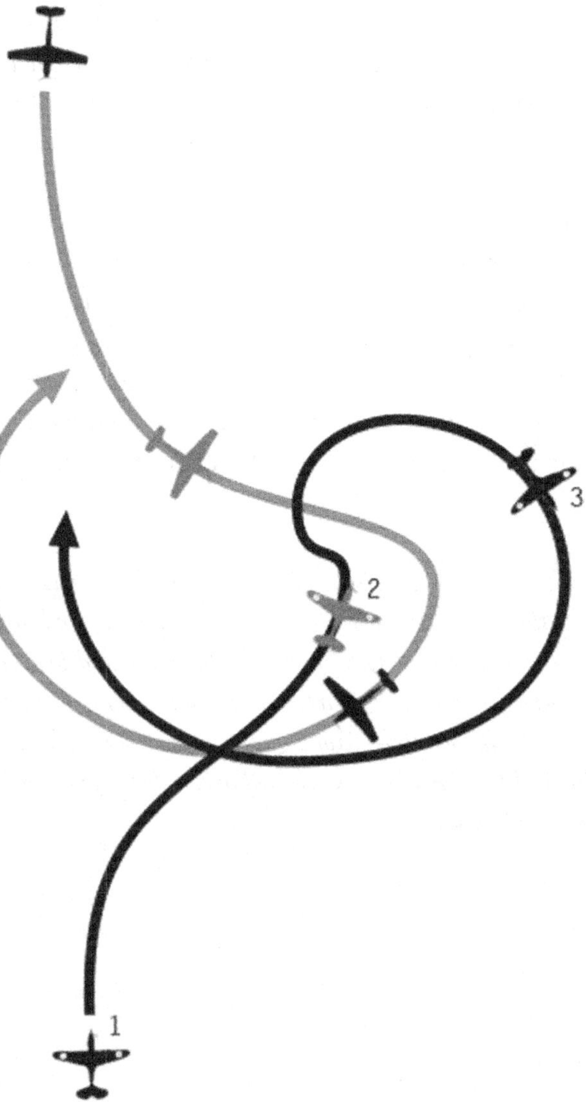

Fig 41. Lead Turn (top view).

The lead turn is in fact a sort of scissor. The objective is to make the enemy commit to one manoeuvre, while you're preparing a diametrically opposed turn to turn the table on him. This works swimmingly in a typical head-on encounter: you KNOW the enemy will go for the shot, i.e. fly with his gunsight right on to you (true pursuit). At time 1, you point your aircraft well to the side of the enemy, forcing him to correct his flight path incrementally to the point of banking hard in the turn direction.

At time 2, he's fully committed and expecting a snapshot at the intersection point. You however, have other plans. Turn towards the bandit and aim your flight path so as to pass him below and to the left (in this example) – to get his guns on you, he must depress his nose (suffering negative G's and a red-out) or very quickly roll to the right and pull hard (the rapidly closing distance won't give him time enough to shoot). You can be pretty sure that the latter will occur, forcing him into a right-hand turn. At that time, you brutally reverse your break to cross his path – preferably in his cold six but just as well with a gravity assist in the vertical, in his high six – by the time he collects his wits you've already completed most of your turn and are working in towards him at time 3. It's like stealing candy from kids.

13.10 HIGH YO-YO

The Yo-Yo is very difficult to explain. It was first perfected by the well-known Chinese fighter pilot Yo-Yo Noritake. He also found it difficult to explain, being quite devoid of English.

Squadron Leader K.G. Holland, RAF

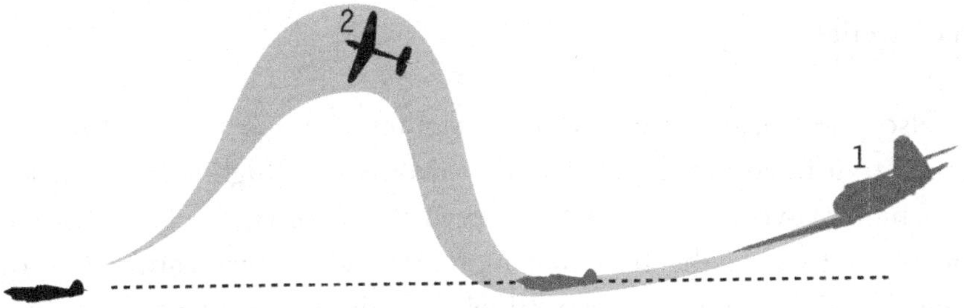

Fig 42. High Yo-Yo.

The high yo-yo is an out-of-plane manoeuvre employed when you need to shed forward speed without sacrificing position or energy – i.e. you momentarily trade speed for altitude and position, only to cash it in shortly thereafter by converting altitude to speed (i.e. the ability to close in). In the picture above, the grey fighter comes in too hot against a curiously unaware defender. Instead of throttling down and pulling hard to curve in behind the bandit – which would cost a significant amount of energy and most likely cause a black-out – he pulls up in a short zoom and hauls in as he's about to cross over and well above the defending fighters track. The cost in energy to pull gravity loads at time 2 is significantly lower than at the juncture shortly after time 1, and allows the attacker to set up a low angle off tracking shot rather than a tricky deflection shot as at time 1.

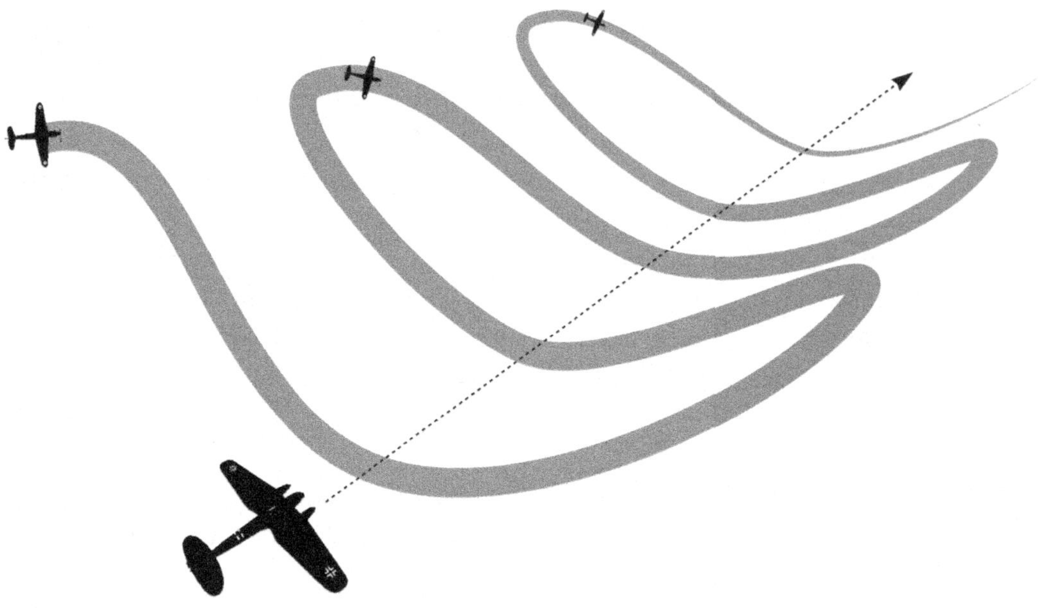

Fig 43. High sides attack (series of yo-yo's).

The high yo-yo is also the basic ingredient in a fighter's attack on a slow and plodding bomber stuck to a fixed heading. This is known as the "high sides" attack - in essence a series of high yo-yo attacks. At the top of the zooms, the target is just behind the wingline so as to give the fighter a high deflection, big planform shot against the target's eight and four o'clock respectively. As the fighter carries through each attack, his velocity vector takes him up high and forward of the bomber to the next attack's starting point. As long as the fighter keeps his energy high and takes care to match his general velocity vector to that of the bomber (rather than getting lost well behind him), he can keep up the attacks as long as necessary – with little risk to himself.

13.11 LOW YO-YO

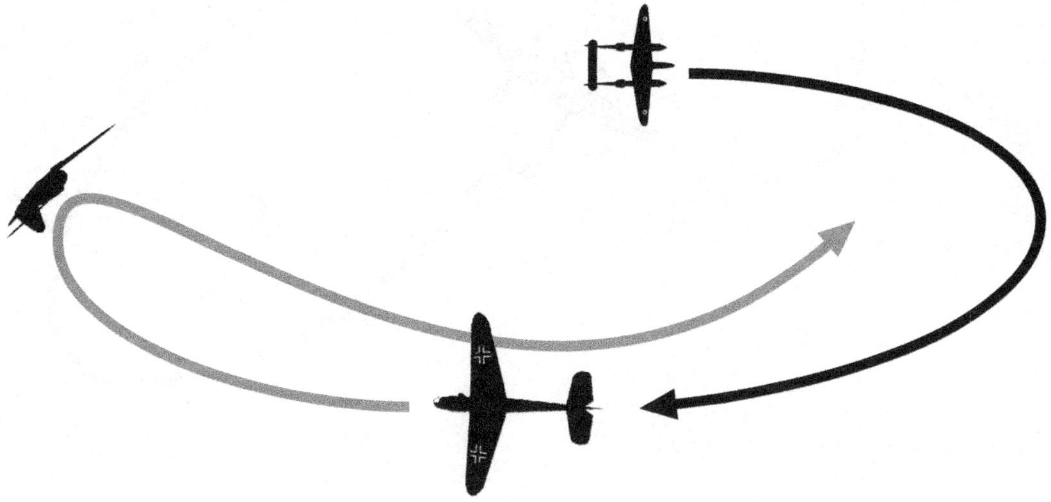

Fig 44. Low Yo-Yo.

The low yo-yo is another out-of-plane manoeuvre which uses a gravity assist to cut across a stagnant flat-circle fight. It requires a bit of altitude, but not much – 500 feet is sufficient, or even 50 for that matter if you need a small energy boost in a stall fight. Following a low yo-yo, it's natural to keep swinging up and down into the enemy's circle, and the fight is likely to progress to a rolling scissors situation. If you're being yo-yo'ed against, the correct course of action is to counter it with a vertical manoeuvre of your own, thus taking the fight into the vertical. To execute the low yo-yo in the turnfight, aggravate your current bank and pull back, then release pressure and roll gently back on the way down. As figure 44 shows, the 109 will be pointing down and to right of the P-38, with the bandit in his high 10 o'clock view, halfway through the manoeuvre prior to pitching up and drawing deflection for the snapshot. The P-38 in this case defends by a low yo-yo of his own, thus increasing his rate across the 109's track and restoring the neutral state. In fig 45 the circles have been displaced and the fighters remain largely neutral, canopy to canopy.

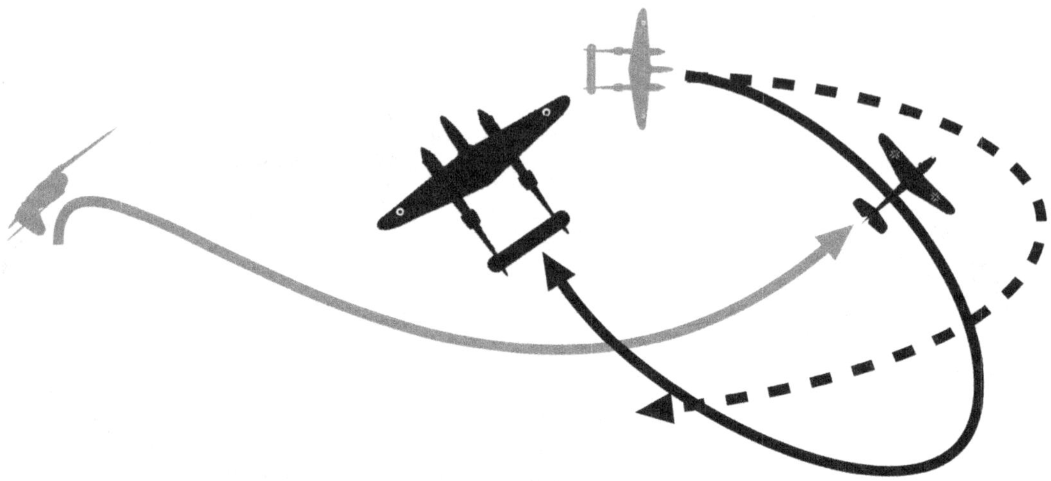

Fig 45. Countering the low yo-yo.

13.12 THE SLICEBACK

The Sliceback is an evil manoeuvre, a dirty trick to pull on a bandit who's comfortably on your long six o'clock. It comprises a gentle zoom with an aggressive skid near the top followed by a nose-low opposite rudder reversal to bring you near head-on with your highly rattled pursuer. You will need quite a bit of separation for this move since you burn a lot of energy in the reversal – don't try it if the bandit is anywhere near guns range. It is best employed by a fighter with high wingloading and considerable rudder authority, in a situation where a flat or oblique turn would most likely give the enemy a full planform snapshot. The manoeuvre requires a good deal of rudder-aileron coordination and should not be attempted without first having perfected the technique. The "trick" is to mask the airspeed decrease with the low speed skid – the follow-up rudder reversal usually comes as a nasty shock.

The sliceback is useful in a standard dogfight as well. If you're energy-rich but angles-poor in a turning fight, the reflexive behaviour is to honk back on the stick and bleed energy to stay with the bandit. This leads without fault to black-out from excessive gravity loads, to loss of visual contact and sacrifice of the energy advantage. A more cautious pilot will normally zoom in the bandit's rear hemisphere and roll his lift vector onto him for a high yo-yo. While this is certainly good in most cases, it's time-consuming and thus allows the bandit a respite during which angles are lost. The outcome is generally a nose-to-nose fight where either or both will succumb to head-on shots or collision.

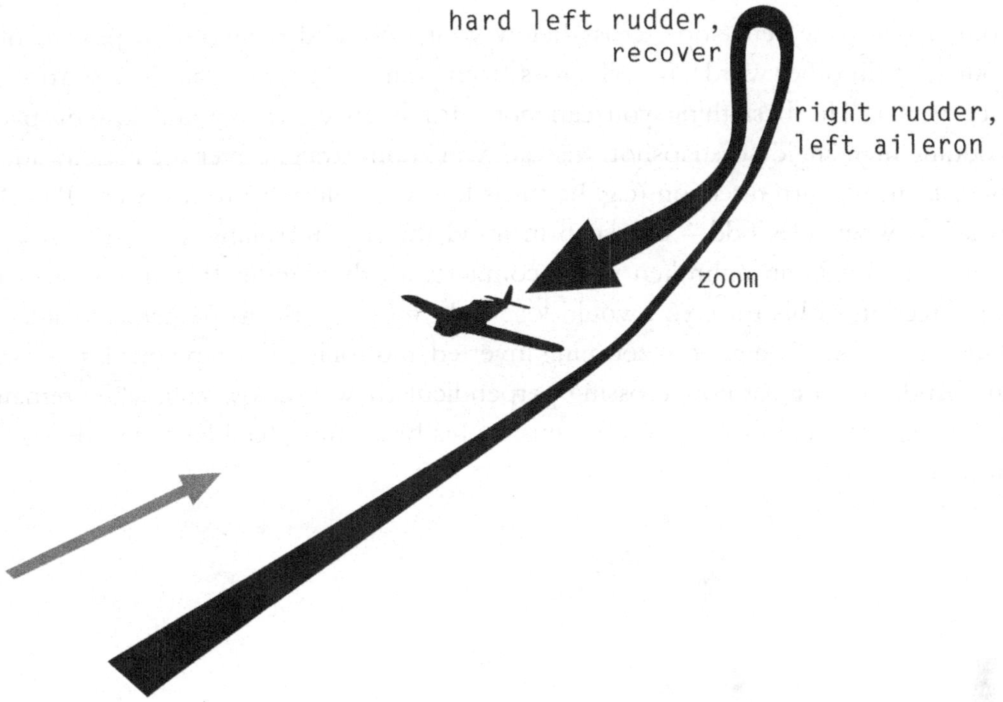

Fig 46. The Sliceback.

In this situation it is far better to use the vertical and a bit of rudder work. E.g. in a left turning fight where you have an energy advantage on a bandit who's breaking hard to left, you nose up and skid right (top rudder) with a bit of left aileron to counter the roll, then rudder hard to left, nose low while adding a bit of right aileron. This is a sort of vertical lag displacement sliceback which doesn't cost you much energy and which gives you a low yo-yo snapshot at the bandit (who sustains his left turn). Follow it up with another high sliceback if he keeps up his turn, and with a steep barrel roll (canopy to bandit) if he reverses.

13.13 LAG DISPLACEMENT ROLL

The lag displacement roll is another advanced manoeuvre to be used in lieu of a high yo-yo when coming in with a ton of smash against a breaking bandit. It's tricky to master at first as it goes somewhat against the grain of reflexive behaviour, but once you recognise the utility of it and the situation in which it can be applied, you'll come to use it without thinking.

As in the sliceback (or pitchback) described earlier, you're coming in hot in the bandit's rear hemisphere at well over corner speed. The bandit spots you coming and starts a sustained turn towards you with a bit of vertical element thrown in for good measure. At the point where he breaks, you can clearly see that he'll move out of your performance envelope, cross below your nose and continue on past to defeat your shot (in other words, he will cross from your low port to your low starboard or vice versa). The best thing you can hope for is an expensive and low-probability skidding high angle off snapshot. Instead, you zoom straight over the badguy and roll away from his turn direction (e.g. he turns left, you roll right) to inverted. This "roll-away" is what feels odd – but keep in mind that you're rolling away from his turn only to maintain an unbroken visual contact with the bandit. If you were to roll in the direction of his turn, you would lose sight momentarily as he passes below your wings and nose. You're now zooming inverted and rolling to keep your lift vector on the bandit, who's just now crossing perpendicularly well below you, while remaining in his high rear hemisphere. As he finishes his break turn, haul back and down, then hack him down.

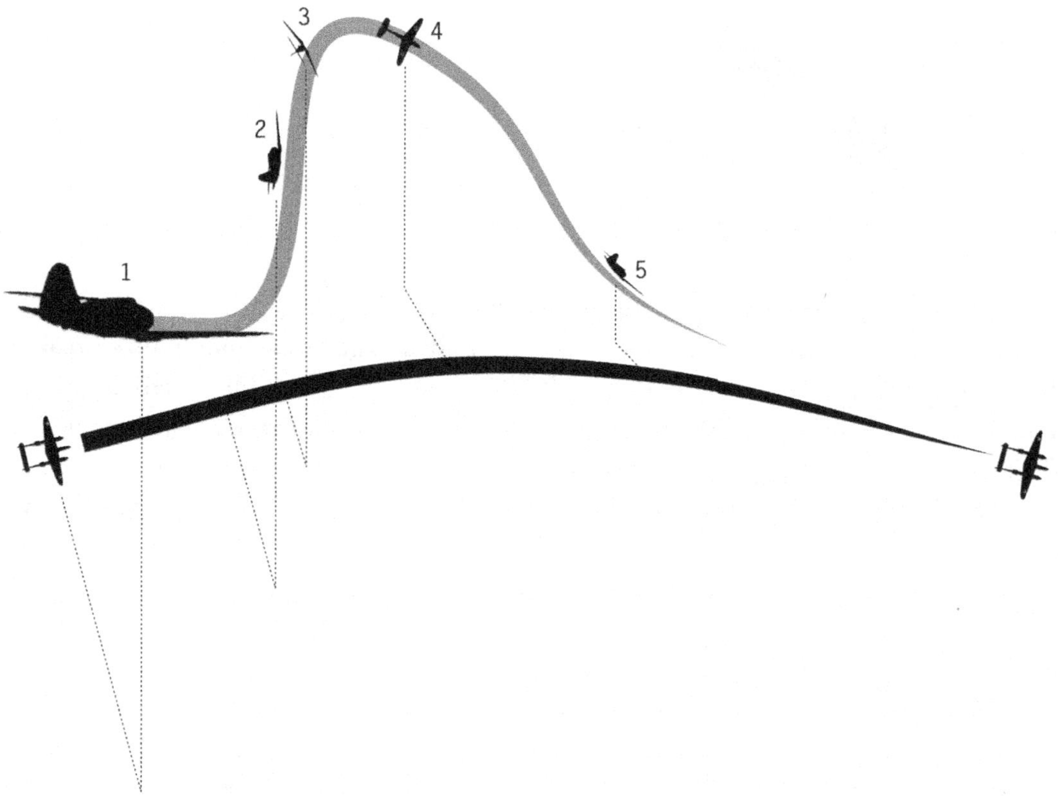

Fig 47. Lag displacement roll. At time 1 the attacker comes in too fast and too high against a turning target. At time 2 he has pitched up and started a roll towards the target but away (perpendicular) from the target's turn direction, in order to keep him in sight. At time 3 the attacker continues the roll to inverted, crossing over the target's path, "floating" as it were. At time 4 the attacker has sailed well past into a high lag position and started his pull down into the target's rear hemisphere. At time 5 the attacker is streaking in for a low deflection

tracking shot. This manoeuvre is different from the high yo-yo pull-up inasmuch as the latter would cause the attacker to lose visual with the target at the critical point (time 2-3).

You can also use this same manoeuvre where the energy disparity isn't so great but where you're fighting a radically breaking bandit who also enjoys a turn rate advantage. As he breaks, pitch up in his rear hemisphere, roll away to keep visual and while continuing the roll to full inverted you tramp down hard on the inside rudder (i.e. closest to the enemy) to slew your nose around while inverted. From there on it's pretty easy to slide down and gun his brains out. E.g. if you have the bandit to your right as you start the manoeuvre and you pitch up and roll right to keep in visual (this being perpendicular to his heading) you will also stomp hard right rudder to slice inside him. If you want to reduce your speed and/or separation in the zoom, give hard top (outside) rudder while quarter-rolled towards the bandit, then quickly give hard down (inside) rudder as you complete the turn to inverted and start rolling toward him. Neutralise, swoop in, gun him down.

Although lacking the conspicuous roll-away, here's a story from an encounter between a P47-C and a FW 190-A4:
As he breaks hard to port, I relax [back pressure on the stick] and zoom a bit for a lag roll, slicing in as he continues the flat left turn. I score a few hits on his high (right) wingtip, skid down and saddle up. Co-E and co-alt, the stall horn blares as I pump flaps and adjust trim to nose high. Ever so slowly, the sight moves through him as I draw lead, then I fire an extended salvo that rips off -sach-'s right wing.

WarBirds, March 21 1998

PART IV: YOU ARE THE ENEMY

CHAPTER 14: PSYCHOLOGY

I went in, hoping to catch him unawares, but did not notice that he was headed straight for me as opposed to away from me, so I gained on him so quick that I had to simply evade his attack to keep from colliding with him. None of his shots during that pass hit, but it did unnerve me a bit.

Reuben D. "accip" Moore, 56th FG(v)

Psychology and morale are strong factors in aerial combat. If you can demoralize your quarry even before the first bullet is fired, you've made the subsequent kill so much easier. Knowing how people react to certain stimuli, and drawing benefit thereof through planned manoeuvring, is a vital part of air-to-air combat.

Every engagement is a contest of wills, a comparison between egos and a measure of morale. The "morale meter" is fickle, dependent on every move, every angle, every pound of energy. While you may start the engagement full of balls and bravado it doesn't take much to sap your fortitude and plunge you into the deepest depths of despair. One false move, one stupid mistake, one bullet even, and suddenly it's the enemy who's on top and brimming with confidence.

Reading the enemy mind, gauging his fear, sensing his desperation and aggression, feeling his hysteria or his coolness, is to me the very essence of online fighting. You don't get that against an artificial intelligence opponent. Online, you can actually reach out and measure your opponent's morale to a tee and characterise him one way or the other: this one's an ace but now he's clearly rattled and becoming sloppy; this one's a grade A rookie without a clue – look at him burn his energy! Look at him stall! Look at him lose sight!; this one's deathly afraid and running hell for leather, he's not even attempting to fight; here's a guy who must have numerical superiority to fight, as soon as he's outmanoeuvred he'll turn tail; this guy doesn't care one whit whether he collides or not; etcetera.

Once you learn to gauge your opponent you also learn to control your own state of mind, to relax and become scientific rather than emotional in your approach to the fight. A fellow who happily runs into any old fight, hell for leather, probably doesn't have that kind of awareness: all he's interested in is a bit of trigger time and yankin' and bankin' action. That guy is an easy mark if you can isolate him. Most opponents are easily intimidated and once you've scared the daylights out of your quarry his critical mistakes won't be long in coming. You'll also learn to discern the leaders and the followers – when you bounce a wingpair, you'll see that one guy is flying slightly ahead and more or less stable whereas the other guy is trailing behind or working hard to format. If their separation doesn't provide a good overwatch already, you can pick and choose whether to kill the leader first or his hapless consort – nailing the leader first will most likely reduce the wingman's morale to a state of absolute panic, whereas excising the wingman first is not likely to lower the leader's morale to the same extent. And then again, it might.

You are the enemy. You are your enemy's enemy, and you're your own enemy. You must *be* the enemy, think like the enemy and feel what the enemy feels in order to defeat him. You must observe him dispassionately, like a vivisectionist observes a lab rat. You, as the enemy's enemy, cannot be afraid, cannot be meek, cannot be weak. You must show an inflexible aggression, an unbending will, an absolute command of the situation. That alone, this intractability, demoralises the enemy and paves the way for his demise. You are your own worst enemy when you falter, when you break, when you hesitate, when you lose heart. Never waver!

Laugh in the face of danger, calm your pounding heart, centre your mind and concentrate on the task. Secure every advantage known to man and nothing, nothing, can stop you. You shall not give the enemy the slightest chance to react or to retaliate, ever. Always on top! Always in command!

Objectively, what are the morale factors, and how do you ensure that it is the enemy who suffers rather than yourself? How do you avoid becoming demoralised if worst comes to worst?

Most fighter pilots are more scared by enemy appearing high behind their wingline than by enemy appearing below and in front of their wingline.
Obviously, pilots fear the immediate threat more than the potential as the former constitutes a higher risk to their well-being. When the enemy is below and in front, you feel reasonably secure and in command. Thus it makes rather good sense from more than one point of view, to attack from the rear with a good overhead of energy. On the other hand this comfortable belief, that contacts to their low front quarter are basically harmless, is an advantage for the fighter who has the considerable amount of energy requisite to make a frontal climbing attack.

Pilots fear formations more than they fear single enemy.
Few pilots relish being outnumbered. While a single bandit with a good head of steam may play havoc with a relatively inflexible formation, at least for a while, it takes either a very capable or a very foolhardy pilot to accept a one-versus-very-many fight. That said, being in a formation is in no way a guarantee for survival (for the formatee) regardless of the odds. Furthermore, the aircraft types involved determine relative morale – a fighter pilot's wet dream is to happen onto a massive, unescorted, bomber formation. "All these targets! Just for me!"

Pilots who are assailed by an unseen enemy are easily rattled.
The first instinct when fired upon, is to panic. This panic, this surprise, causes an instant loss of confidence – "how could I fail to notice the threat?", "where on Earth did he come from?" – and lends, right or wrong, an aura of invincibility and untouchable superiority to the enemy. If the blow is not mortal, the stricken pilot must now not only struggle to get out of a sticky situation but struggle to regain a semblance of confidence as well. Surviving the bounce may prove to be the requisite shot in the arm, and then again it may not. In any case, the defending pilot has an uphill struggle, and, if repeatedly bounced or assailed without having seen his assailant, his morale will plummet to rock bottom sooner than his sorry crate hits the earth.

Pilots who are outmanoeuvred or stall repeatedly lose faith in their own capacity and in their craft.
To realise that you're not master of your stead is an incredibly demoralising discovery. Is it the aircraft which lets you down, or is it the pilot who mismanages the aircraft? Most of the time the pilot is to blame, yet the pilot likes to blame the crate. He will curse and sweat, damning his machine and using every foul word in his vocabulary as he struggles to bolster his faltering heart. Repeated failure to close with the enemy, gunnery which never improves, loss of energy and seemingly untouchable enemy aircraft all contribute to lower the pilot's morale and make his moves shoddy, clumsy and without finesse. A pilot who doesn't know the limits of his proficiency and the envelope of his aircraft is most conspicuous in the air and an easy mark for the discerning eye.

The pilot who scores the first hits, irregardless of damage, establishes a moral superiority.
Being hit and not knowing whether the hits were critical or not, especially in a turning contest where you were reasonably certain of not being hit, is decidedly unhinging. While being hit probably won't alter your manoeuvring at that point – unless the hits are persistent or their origin obvious – it nevertheless swings the morale meter in favour of the enemy. He has the audacity and the skill to land hits, on you! And here you are, struggling to keep him in view! If the hits register well and cause your crate to lose performance, your morale will surely hit a new low water mark. At that point many pilots simply give up. Others grit their teeth yet harder and adapt, swearing to all that is holy, and to some who aren't, that they – will – not – succumb! The shooter on the other hand, may just relax his vigilance the critical amount upon seeing that his fire has crippled or severely wounded the enemy. Never relax! The show isn't over until the fat lady sings.

Losing sight of the enemy is demoralising.

When you're engaged in an especially taxing fight which requires you to keep visual contact with one or several enemy at a time, where you must ever keep rolling, diving, zooming to stay out of lethal volumes of fire, if then you lose sight of the enemy known to be within easy reach of your unprotected astern, you will rapidly become demoralised, to the point of panic. Now you see him, now you don't. Now you hear his fire slamming into you. Where is he! Roll, yank, dive! There he is! Roll again, break! Where is he! I can't see him! Where is he!!! Naturally, your panic and your inability to spot the enemy affects your ability to manoeuvre profitably against him – and you will shortly find yourself sitting in a burning, plummeting, hulk – unless you immediately manoeuvre so as to maximise your chances of picking him up visually. In the following story, morale features heavily in the combat:

I sweep southeast to east and end up near Echternach where I spot two 109s climbing out on the beeline to the front. I have at least 2km on them. Alas, my attack is completely disjointed by bad trim and a fumbling approach. I streak past the no doubt startled trailer with but feet to spare, without having fired my guns. I jink a bit and continue towards the leader some 2000 m ahead. I'm not bothered by the guy I left but reckon he'll warn the bloke ahead. Verily, at the last moment he goes into a panicky guns defence. I blow past, guns silent still, and extend southeast. Bad karma. I slap myself. Glancing back, I see them both give chase. No surprise there.

I shed some altitude to gain as much separation from the track as possible and lead them away to the southeast. They retain their 2000 m separation. When the leader is near guns range, I do a sloppy rolling scissor, lose sight and fly uncoordinated through a couple of rolls, straining to regain visual. Man I suck! However, they suck too - the trailer manages to lose sight of me somehow, leaving me alone with the leader. I take him through a series of easy lead turns, noting that he doesn't seem shy to stay in the fight nor adverse to burning his smash. A couple of 7,62 splatter my empennage in a badly timed lead turn but they don't rattle me. From then on it's my fight. I draw lead and pepper him, letting him settle fully into a state of panic. He tries to extend, still in guns range but out of cannon range. He manoeuvres heavily as I goad him to evade, setting himself up for several full planform shots. I insert a few 20 mm in his mid fuselage and continue to hose him down with the popguns. I can feel his performance degrade sharply, and subject him to a few high yo-yo's. He's right out of options now, turning desperately for his life. I've seen it coming and watch him stall out low in a left turn, decorating the landscape with a pretty plume of oily smoke. I cut RPM's, throttle back and hug the deck down to Metz, happy that I didn't mess up altogether.

WWII Online, December 30 2003

Losing wingmen or flight members is demoralizing.

Loss of a wingman or flight member is a personal blow. It reduces your numbers and creates a feeling of despair, however fleeting. You must ask yourself whether you were party to the loss somehow: did you fail in calling out a threat; could you have manoeuvred to help; were your instructions to the lost pilot lacking in some way? And the assailant – does he have position to cause yet more loss? Will you be able to return to base at all, or will you all die miserably deep in enemy territory? Succumbing to such thoughts in the cockpit is devastating! Shrug off the loss and keep fighting!

Being shot down repeatedly over your own airfield is demoralising.
Your own base should be a safe haven, right? What audacity then on part of the enemy to loiter nearby and shoot down countless friendlies shortly after takeoff – some even before their wheels retract! You of course see the threat, at least after having been shot down twice already, and wow to avenge your mates. As you claw for altitude, firing your guns in frustration at the rapidly disappearing enemy, you are creamed by yet another enemy streaking in on your six! Have that happen to you a couple of times and you will be fit to throw up at your own ineptitude. Avoid such deaths at all cost, but let it happen to the enemy as often as you can.

It is demoralising to have an enemy follow you around, him clearly in a good guns position yet waiting for a REALLY good shot opportunity.
Being unable to outmanoeuvre an enemy is highly irritating but not always directly lethal. Not being able to shake an enemy off your six though, is something else. For every move you make, he has an answer. Whatever you try, he's right there on top of you. Even scissors doesn't work. He anticipates your every move and isn't foxed by desperate attempts such as cutting throttle or throwing out the boards (flaps). He just won't let go. However, he's not firing. You've been dodging him for a good five minutes already, and he has yet to fire his guns? Is he out of ammunition? Why then is he following you around? Is he trying to make you auger in, or run you out of fuel? He keeps coming closer. You can see the individual blades in his propeller, the evil grin beneath those blank goggles… NOW he fires, when he cannot miss! You dodge, you skid, you roll like crazy. Still he won't let go. You swirl like a dervish, and he matches your every move – not wasting a single bullet! YOU are sweating bullets! You know that the instant you fumble, the instant you mush, the instant you become predictable, it will be the end.

Next time you go up, you know your nemesis is around there somewhere. You hope to never meet him again.

It is demoralising to lose altitude and position for no apparent gain.
Enroute to your primary target you are engaged by an enemy attacking from superior altitude. He forces you onto the defensive and deflects you from your mission. While

your skill keeps you alive, the enemy effectively bars you from levelling the energy disparity and forces you further and further away from completing your mission. At long last you shake him off, but now you're well below the average energy level for the area and you've wasted precious fuel too. You know that if you press on to your objective you will be confronted by numerous enemy all enjoying a substantial energy advantage. So you climb again, only to be deflected yet again by more enemy. Disgusted and dejected, you break off your mission and head home with bingo fuel. The enemy force as a whole takes on an aura of apparent superiority, causing you to regard them with increased respect. Clearly your best shot is to spend yet more time climbing and amassing a greater force to deal with them, something which is both tedious and connected with a lot of effort.

ALL IS NOT WHAT IT SEEMS TO BE.

"All warfare is based upon deception."
Sun Tzu, The Art of War

In the individual fight, morale and psychology factors are at a premium. Take for instance the situation where you're extending from a bandit-infested area with one of them on your long six: to appearance, you are fleeing the scene and your pursuer, sensing your fear, thereby gains a definite feeling of superiority. After all he's on your six with a potential guns situation, right? He has good closure too, you can almost hear him lick his chops, hear him gleefully report to his mates that he's chasing one sorry fornicator and expect to kill him shortly. His morale is so great, his superiority so evident, that he believes nothing will shake it. Well, he's in for a nasty surprise. For while you initially did depart the scene in something of a hurry and with depleted SA, you have now had time to take stock of the situation: the cloud of enemy is far behind, wheeling and looping still and not in the least a threat to you unless this single bandit calls for assistance – which is unlikely, after all he's in command and he doesn't like to share his kills – you are extending deep into enemy territory far away from likely flight corridors and the bandit on your tail is only closing slowly. Giving you plenty of time to consider your next move. You are calm and collected. You've extended in a more or less straight path, jinking every now and then to check your six and to keep your pursuer in visual contact, measuring his progress. This alone makes the enemy believe that he can safely truck up to comfortable guns range where you will obligingly play sitting duck – not! As he draws almost into guns range, you execute a violent oblique spiral or some other creative means of transition to the attack, defeating the bandit's shot and swooping in on his tail in one fell manoeuvre. The hapless bastard makes it even easier for you by keeping his lift vector on you, blowing great chunks of energy for no gain whatsoever. Now you are on his six, the tables are turned. The enemy, annoyed at first, now kisses his morale goodbye – he panics. The easy mark has turned out to be his worst nightmare, he must get away! Away! Too bad he has burned all his smash already, and you're on his six. Sensing disaster, the enemy makes a feeble turn away, homeward, desperately nosing down to gain speed. Too bad the ground is already quite close. He trusts his acceleration, but no crate can draw out of guns range that fast. A withering hail envelops him, drawing oil and coolant leaks, puncturing his fuel tank, wounding his crate. He wriggles and waggles, to no avail, and then he litters the landscape. "But I was on his six! How the blue blazes did he do that!"

Play the enemy. Lure him along. Give him false impressions and then show him what is real, when he has no chance of redemption. Smoke and mirrors.

CHAPTER 15: AGGRESSION

"We're going to kick the hell out of them all the time, and we're going to go through them like crap through a goose."

General George S. Patton

The fighter pilot must possess a singular will to close with the enemy. An aggressive spirit, one that doesn't recognise defeat, that doesn't acknowledge the remotest possibility of defeat, goes a long way in combat. Naturally, aggression alone won't defeat the enemy – it must be coupled with cold calculation, a heightened and ever awake awareness, with determination and insight. Merely being aggressive and clawing straight into the teeth of adversity will most likely get you killed in a hurry. What distinguishes the successful pilot is his ability to temper his aggression when necessary and his ability to unleash it to its full fury at the most auspicious moment, without allowing it to blind him.

Once you start feeling defensive, outflown, boxed in or pushed hither and thither by the enemy, your aggression takes a hit and causes you to fly defensively, predictably, benignly. At that point, or rather some time before that, you must work up a storm of adrenaline, a holy wrath, and make your every move devastating, calculated, inveterate. Bare your teeth, curse the enemy and kill your despair. You must be deliberate, full of purpose and acutely aware of the slightest shift of advantages. Did the enemy miss his shot? Does he not use his roll rate? Wasn't he a bit slow there? Did he not fail to press his advantage in that instance? Isn't he stalling now? Look for every opening, jump on his every mistake and regain your confidence little by little. The victor is not the guy who flies the best but the one who makes the less mistakes, and every pilot makes mistakes.

A careful, defensive approach only serves to give the enemy a shot in the arm. When it is you who hold the initiative, don't hesitate! Exploit it without compunction, without holding back. Overwhelm your enemy and give him no quarter for he surely will give you none.

Adrenaline in particular is most important. You must want to win with every fibre of your body. You must feel righteous anger boil through your veins – don't be a sap! Don't just sit there, manipulating the stick as were it made of porcelain, snug in your little chair. Fly like you've never flown before! You are a FIGHTER! Concentrate on the task, let your moves flow naturally and KILL! KILL! KILL!

That said, naked aggression bluntly projected and pressed to the hilt can land you in a world of hurt too, one of your own making. Never let your aggression carry you over the top so that you disregard the basic tenets of air combat. For instance, pressing a diving attack beyond the point of profitability only serves to bleed your energy. If the enemy breaks away or otherwise slips out of your guns solution while you're still streaking down on him, making it obvious that you must pull excessive G's in order to hit, immediately break off and regain your superior position. Pressing the attack at that point, or worse, braking and trying to reel him in, will diminish your energy and make the whole combat immeasurably more difficult. Be aggressive, but temper your aggression. Too much aggression can cause you to take on a bandit too many, and even if you feel confident in your ability to handle both, you're getting bogged down and predictable. Before long the two "easy" bandits are joined by an expert or two, and a couple of more "easy" bandits, making the whole situation quite untenable. Know your limits! Similarly, aggressively pursuing an enemy outside guns range puts you in a predictable and unproductive situation. If you cannot kill him outright, let him go and look for more obliging targets.

Aggression is not limited to one-on-one situations, it also applies to the combat zone at large. Ask yourself, which side is the more aggressive in general, who has the initiative? Can you blunt the enemy drive by picking them off early, can you deflect them from their mission, can you single-handedly turn the tide simply by posing a threat? Is the enemy confident, desperate, relaxed, disorganised? Before taking off you should consider the combat zone as a whole and seek to maximise your effect on it. In other words, what is going on in the online world and where will you be of the most use?

Generally speaking, the participants in the online "war" tend to flock towards the areas where the most action is to be had – the frontline, or what passes for a frontline. This flocking follows a Poisson-like distribution and can be regarded as a force unto itself really, a self-propagating meatgrinder as it were. In this frontline you will find innumerable small and large dogfights, or furballs in pilotese, most of them at or very near the deck. As fresh fighters appear they are inexorably drawn down and into these churning furballs where they, for the most part, are chewed up and spat out in teensy bits and pieces. Few who participate in this sort of combat ever makes it home, again testament to the low regard for the virtual life. The side which manages to feed in the most fighters, or draws the most benefit from automated and player-manned ground fire, will for the most part keep the advantage and mangle their way forward. Diving singly into such furfests is seldom profitable, yet such practice is by far the most common for the simple reason that pilots are impatient and generally seek instant action: trigger time and yankin' and bankin' to their hearts desire. Personally I find this practice incredibly silly and repetitive, and not in the least gratifying, but that's just me.

In such a setting, aggression, on a large scale, is twofold: you have the aggressive grinder churning its way across the landscape, and you have the aggressive interdiction of aircraft heading to the grinder. While the former is of the hyperactive and brawny sort, the latter has a studious and selective character. The furball requires little thought or expertise beyond that of following a crowd whereas the interdiction-based fight requires knowledge of and adherence to bona fide tactics. Needless to say, the fighter who's looking for a quality fight is more likely to look beyond the furball and seek to project his aggression in a controllable environment rather than the chaotic one. This controllable environment is habitually found in the "cruise corridors" between enemy airfields and the most likely enemy mission targets. Patrolling these corridors and loitering near enemy airfields is seldom uneventful, and you get to practice your SA and energy tactics with a lot less risk to yourself. If you're sufficiently good at chipping away on the enemy thoroughfare, they will eventually wisen up and send someone after you, making for some quite interesting fights. If you're feeling particularly aggressive, bring along a bevy of friends and make yourself comfortable near an enemy airfield: with employment of proper team tactics you can make life of the enemy exceedingly frustrating and strangle the flow from that particular field, temporarily lifting the frontline siege. Here's a short tale to illustrate the method, and its superiority over the "brawn-fight":

Appearing over the Abbeville climbout area again, now at angels nine, I spot a bandit low on my ten o'clock. I go in and hit the climbing 109 full and square in the port wingroot, correcting some with port rudder to keep fire concentrated. -hahn- blows up as I tromp starboard rudder to clear the burning hulk. I exit the area to the west for fresh altitude. Returning to my hunting grounds, I find yet another climber out of Abbeville and dive down into his cold six. An FW. Frrrrrt boom... lipfer buys the farm. Extend. A little later, the exact fate is experienced by another Messer, driven by konrad, who fails to pick up my Mustang swooping down at his low six. Boom! Homeward bound, I enter a fight against an FW who's running back to France. I land a few hits but since he's all awake and rolling hard, I let go as I feel my SA become compromised. I earn an assist when grendl turn his FW into scrap metal a while later. I earn another assist on -hahn-'s 109 in another scrap, then set course for Blighty.

WarBirds, October 25 1998

Aggression comes in many forms and shapes. Consider Erich Hartmann, arguably the world's most successful fighter pilot who in little over two years fighting scored 352 confirmed victories. Without knowing his story, one might think that he'd be a singularly aggressive person with almost magical powers to dive in and sort out 50 plus bandits in one swoop. Not so. He was a coolly calculating killer, an airborne assassin possessing superior situational awareness and abiding by an extremely rigorous engagement doctrine. His "dicta" was an incredible simple formula which every aspiring pilot should take to heart: See - Decide - Attack - Coffee Break.

SEE. He who spots the enemy first scores a major advantage, allowing him to manoeuvre profitably while the opponent trudges along in blissful ignorance. 90% of all kills are made against pilots who never saw the threat. Thus, keen eyesight, all-round vigilance and the ability to see the enemy at extreme range, is of utmost importance.

DECIDE. Is it safe to attack? Can you get away with it or are there factors that should induce caution? Can you attack from your current position or must you manoeuvre? What will your action be after the attack? When is the optimum time to attack?

ATTACK. Make your attack swift and merciless at the most auspicious moment. Fly close to your victim, from dead astern or slightly below if possible, and shoot only when you're certain to score an immediate kill. Close to minimum distance - your sight should be black with the enemy.

COFFEE BREAK. If you cannot attack safely or without the enemy taking drastic action to evade you, take a coffee break. I.e. disengage and look for an easier victim. If you do attack, make it in one single devastating pass and immediately disengage to a safe altitude or a safe area to regain situational awareness.

What all this boils down to is a fighter who is not merely extremely aggressive at the proper times, extremely cunning and cold-blooded but also exemplarily cautious and security-minded. He displays an abundance of discipline, knowing full well when to engage and when not to, the possibilities and potential effects of every last action. Erich Hartmann epitomizes the perfect killing machine, and should do well as a role model for any up-and-coming online pilot.

Fairness, or entering a fight on even terms, or flying as the enemy expects you to, is entirely out of place in the air. If you are to succeed you must stack as many unfair advantages as possible in your favour and never hesitate to use them. Altitude, speed, surprise, friendly fighters, team tactics – use them all! Only a fool gives the enemy a chance to hit back. You must seek out the enemy where he is the weakest, where he has no room to manoeuvre, where you outnumber him and can bring every conceivable advantage to bear with maximum force. It's a grim business, but you must do it to him before he does it to you.

CHAPTER 16: THINKING AHEAD

In line with the "big energy" picture outlined previously and with your knowledge of where the frontline is drawn and the likely routes to and fro likely hot spots, it is quite easy to conjure a mental four-dimensional picture of the sky at large. You need to keep this mental picture in mind at all times, and be acutely aware of noticeable flows and counterflows of advantage and position. In that regard, it doesn't take more than a few fighters to "put a lid" on an area and shift the focus from one place to another, or for one desperately disengaging fighter to drag along a whole gaggle of enemy fighters to his home plate, where they will pose somewhat of a discomfort for friends just becoming airborne. The successful fighter maintains this wide-area awareness at all times, even when engaged in single combat. He makes careful note of the altitudes and numbers and flight profiles of friendly aircraft, he notes how many of them who return and in what condition, he takes similar stock of the enemy of course and always keep in mind his position and altitude relative to enemy bases and routes. If he did not, he would always be fighting against surprise and disadvantage.

Thinking ahead and anticipating the moves of your enemy is applicable to single combat too of course, and has much to do with psychology. Let me relate a little story which highlights this. I was out on my standard mission of intercepting the outbound enemy lanes, swooping enemy takeoff areas as it were, and came upon your ordinary SA-challenged Ju-87 Stuka puttering out to the front:

I swept down from 3500 meters and left him a burning wreck in a single pass, clawed up and perpendicularly away from the outbound corridor to regain my perch. Figuring the Stuka jock would be rather miffed at this inopportune and early ending of his sortie and thus likely to jump into a fighter to even out the score, I made my way back to the enemy field to find him before he got too far away. Quite as expected, a single bandit was seen climbing out towards the very same quarter of the sky, no doubt expecting to see his assailant patiently awaiting his speedy return. He didn't appear to check his high six very diligently – no sudden rolls, no deviations of heading – so I swept in and disillusioned him yet again.

Five high-explosive cannon shells sent him packing into the ground. Now, having been bushwhacked twice, I reckoned the enemy pilot would be fairly steaming and not quite as likely to repeat his past mistakes from the very same spot, especially since there were no friends of his to help him along or call his six. Another enemy field not far remote would most likely be his next choice, wherefrom he might get some altitude and come looking for the pesky swooper who had made his day so bothersome. All of this went through my head as I zoomed back to altitude, now at greater speed and less angle to cover the intervening distance to the other enemy field.

You may appreciate my satisfaction when I encountered a bogey on my low 1 o'clock making best time toward the original perch. I cleaned up (nose down for speed, trim tabs forward) and let him move past and below my nose some 4000 meters distant. He was no more than a speck at this time. Then I swept in in a lazy lead pursuit curve, coming in with good closure on his low seven o'clock where his stern view was the worst. I half expected him to check his six at this point, to make it more interesting, alas he didn't. Closing in to 150 meters, I clobbered him with a judicious amount of cannon and machinegun fire, again sending him earthward trailing black smoke and flame. I then returned to base.

While this all sounds frightfully self-indulgent and uninteresting, and perhaps it is, it nevertheless serves to highlight the need for putting yourself into the enemy's head and acting accordingly. Following the first kill, I could very well have loitered and made myself more conspicuous, and after the second I could very well have remained at my original hunting ground – but I didn't. I reckoned the enemy would be looking for me, that he would be bent on revenge, that he would act in a predictably human manner, and he did.

You get all kinds of clues in the air which you must process and use as the basis for your decisions. Predictability of behaviour and action-reaction is not hard to gauge, and should be what governs your behaviour and flight profile at all times.

Tracers are dead give-aways in the sky, as are smoke plumes and various spilled fluids. Take due note, figure out whether they are friendly or enemy based on their headings, and move to intercept. Since a crippled aircraft attracts scavengers like nothing else, be on the lookout regardless whether you're the one helping a friendly or finishing off an enemy, or flying cover for those below.

"Thinking ahead" is always a factor. Consider the situation when you and your wingman happen upon a low bandit: rather than going in more or less at the same time, with but a few hundred meters separation, thus effectively fighting as one, think ahead! If the first fighter misses his shot or doesn't deliver a killing blow, the target is highly likely to make an immediate guns defence break. The second fighter is then faced with a particularly tricky shot for which he is ill prepared – he simply doesn't have the requisite reaction time or the AOA to line up for a good shot. And if the first fighter does deliver a killing blow, the second fighter is redundant anyway. The better practice is to open up separation between the two fighters in the attack, to about 10-15 seconds or about 1000-1500 yards, by having the second fighter "hang back" at the point of winging into the attack. The net effect of this is, in the case the first fighter botches his pass, that the second fighter will be streaking in on the bandit just as he recovers from his fright and is pointing his crate at the rapidly disappearing assailant. Ergo, a simple shot for the second fighter. And should the first fighter succeed in his pass, the separation is easily closed by cutting across the post-attack disengagement curve.

Thinking ahead is also in evidence when you're hanging about a particular killing zone, such as an enemy airfield, ground feature or part of a traffic lane. If you loiter in any one area for too long, and particularly if you're obnoxious enough to down a large quantity of enemy there, you're running a definite risk of overstaying your welcome. You're fast becoming predictable and will eventually be faced by enemy who have amassed numbers and altitude to deal with you. Thus, if you find a particularly sweet spot, don't make the mistake of thinking that it will last forever. Make a few kills and then relocate in anticipation of the enemy response. Take the little story I told earlier in this chapter as an example of how to avoid overstaying your welcome.

A typical rookie mistake is to fight in the present without having a real idea or plan of what to do other than what is imminently possible or dictated by reflex. Thus the rookie points his crate at the enemy at all times ("hey, flying towards the bandit is aggressive, right?") and always works to fly the shortest route to the enemy regardless of cost and practicability. Consider the guy who, having orbited an enemy field for several agonising and frustrating minutes without contact (heh), spots a bandit on the runway and immediately swoops down to gun him down – ignoring the fact that the enemy field is literally bristling with hungry barrels or that he's likely to lawndart from a combination of excessive speed and target fixation. Or consider the guy who flies towards his opponent, then to turn expensively around to face him again while the wily foe seemingly effortlessly dances around him outside his grasp, plays him like a bullfighter plays a bull, until he's out of breath and out of options.

When entering a fight you should have your next two-three moves plotted out in advance, and keep plotting the next few moves while you're executing any one of them. Thus you play the enemy, force him to adjust to your game plan, and, in the unlikely case that the enemy is leading you about, you will know who's in possession of the initiative. This way of fighting pertains both to the offensive and the defensive alike. Let me give a few examples:

Approaching the enemy head-on I decide to double-fox him, then to work in the vertical. Thus I point well beside him, forcing him to adjust to my decision of how to run the fight. As he turns towards me, I turn into him and beyond him, again forcing him to comply. I see him roll to match my new heading. Now we're getting close. He's gunning for a tracking head-on shot, just as expected. I won't give it to him. Just outside effective range, I roll to reverse across his nose and pitch up a little bit to get well outside his plane of manoeuvre. His shot ruined, he has no choice but to roll and pull his lift vector onto me. We're well past each other now and I know exactly where he's going – he'll be pulling hard to the left to come around after me, because I passed him on his left. I know I have him already, from the moment I saw him try to track a shot. As we merge, I zoom and roll again to pitch up right above him while he's still in the process of collecting his wits. I see him turn below me. I'm waiting for the right moment. Will he lose sight completely and go straight, or will he regain visual and zoom after me? If the former, all I do is pull down on him and gun him down. If the latter, the fight will continue in a rather more gratifying manner. So he sees me now and zooms. My energy advantage is now 2:1 or greater depending on how many G's he pulls in the reversal and initial zoom. So, I can afford to go from my floating position to a lazily spiralling one. I quarter-roll and put him just in front of my wingline, quite near the wingtip, and climb gently without losing too much airspeed. This is the rope-a-dope, and he's buying it. And so it goes – more or less according to plan.

Decisions must come swiftly. High bandit, crossing from 11 to 5 o'clock. Pretty slow. I have energy. Altitude difference is acceptable. Low front aspect shot, reverse for seconds at co-E state. Execute.

Furball, unknown composition. Wildcards from east, all altitudes. Check out top layer, clear out approaches out to 2-3 minutes flight time. All clear. Blow through furball to west, take any shot, chandelle and repeat. After second run, go back to clear out high east approaches. Execute.

Bandit approaching high nine, looking me over. Turn slightly to keep in vis and proceed as if unobserved. Play sitting duck and surprise him. Here he comes, get set. Wait for it. Wait. Wait. Now - barrel roll!

Bogey low 10 o'clock, opposite heading, level. Good speed, non-manoeuvring. In in in! High seven run, convert quickly to low seven, gentle zoom past the shot, roll to float.

Bad shot. Release, disengage.

He tags along. No closure. Pull aside, then chandelle if he persists.

Bad move. Lost vis. Regain! Extend to clean up and start over. He pursues – good! Moderate closure. Plenty of time. Scissor time, set him up. He cuts across, increasing his closure – good! Here we go, pitch and roll! Guns guns guns!

Talk to yourself. State your intent, comment your actions and map out your choices, in real time. Talking to yourself may seem daft but it's a great way to learn and to become aware of the choices you make, and, ultimately, the errors you make.

CHAPTER 17: COMMON SITUATIONS

Although every engagement is unique with regard to energy, angles, detection, numbers, skills and so forth, it is nonetheless possible to identify certain situations as commonly recurring. In this section, we explore some of the many possibilities open to the single pilot encountering another single pilot of equal skill, in situations that tend to pose the greatest problems for fledgling pilots.

17.1 THE MERGE

We've discussed the level head-on merge already to some degree in previous chapters, but there are more ways to deal with it than the lead turn. On the whole though, I venture to propose that the lead turn will nearly always be successful (it haven't failed me yet!), especially with regard to avoiding the sorry spectacle of head-on crapshoots and the risk of collision. If YOU avoid the collision, it won't happen. Simple as that. The merge variants discussed here presupposes that the opponents both choose to fight.

The level merge is somewhat of a chicken race even when you're eschewing the head-on run. The question is who will make the first move, who will show his hand first? In the case of the lead turn, it appears as if the lead turning fighter is acting defensively whereas in reality he's only allowing the enemy to set himself up for a quick kill. The same is true for other gambits as well, but in those situations observation and reaction is at a much higher premium than in the leisurely planned lead turn.

As a rule, the post-merge manoeuvre is rarely if ever played out in the vertical going down (Split-S and subsequent zoom). Such a move defies logic, particularly when the merge is at or over corner velocity: it sacrifices altitude while making the reversal extremely expensive in terms of G forces and turn radius due to the high speeds. Similarly, a flat turn reversal is also a rare occurrence except when the fighters are merging at speeds which rules out vertical moves. Should your opponent be such a fool as to reverse in the horizontal you will naturally swoop in on his six immediately by way of a vertical or oblique reversal – the gravity assist allows you to cut him off.

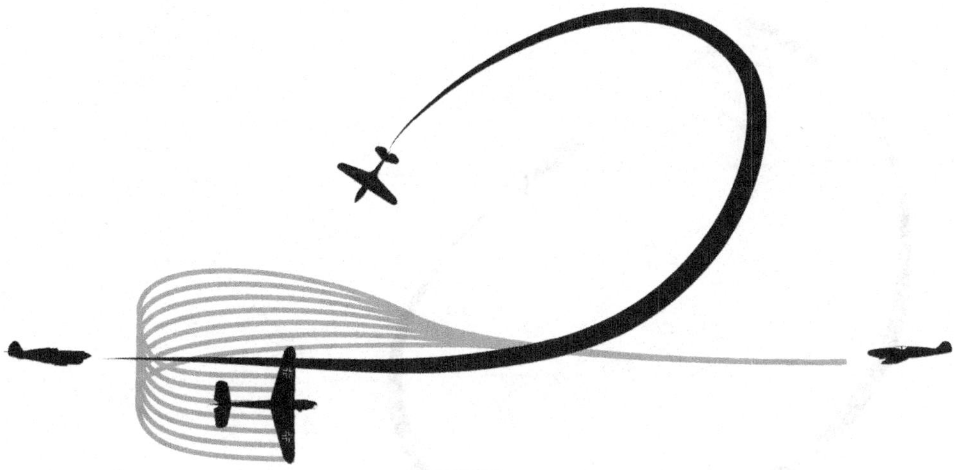

Fig 48. Vertical reversal against a flat-turning bandit.

The most frequent merge manoeuvre is the vertical or oblique reversal. If you both haul in and stick your lift vectors at each other, the result is an inverted head-on pass. That's a big no-no. It is here that the chicken race is manifest – who will commit to hauling in first, and what are the possible responses? The trick is to observe the enemy closely and make a split second decision based on his likely and apparent action, right before the merge and right after the merge. This is also dependent on your relative position in the run-in to the merge, your attitude at the merge and your "advertised" action. An important thing to remember is the urge to fight "upright". Most pilots dislike the inverted second merge, instead they struggle to half-roll to fight the enemy "sunny side up". Here's opportunity staring you right in the face: eschew the half-roll and go straight for the kill while inverted. You will get more time in the gunsight for a steady, unloaded shot whereas your opponent will barely have righted his crate before it's time to get down to guns.

Personally, the times when I don't use the lead turn, I usually go for a low merge as in figure 49. This means that I nose down to pick up as much speed as possible (if it's a short range encounter) and pass close to the bandit below his port wing in a slightly banked state, preparatory to zooming. This forces the opponent to look forward-down and roll a tad to maintain visual, keeping his velocity vector forward-down for a critical instant, and usually results in myself getting a slight head start in the reversal. In addition, thus I can expect to maintain visual after the merge in my high rear as I initiate the reversal. Passing above him would mean that he would automatically keep his lift vector up at the merge, thus inducing him to turn after me without delay. That fraction of a second gained in initiating the reversal is usually all it takes to gain angles for an inverted canopy shot on the bandit.

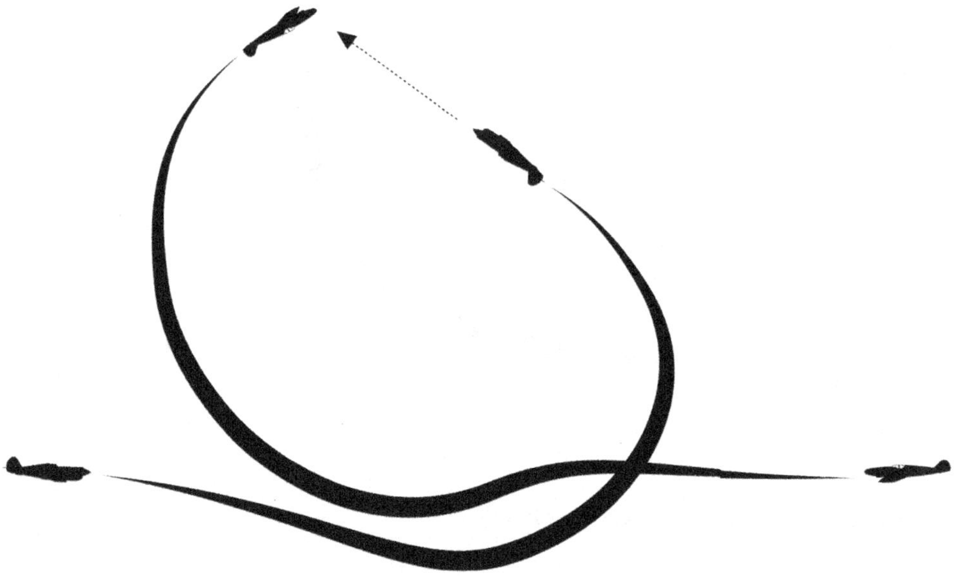

Fig 49. Low merge, vertical reversal.

What happens after that is largely up to what the bandit does: if he's slow on the reversal, but does reverse, I'm already on his six with a purely vertical or oblique move. Should he haul around steeply, I will most likely haul around too after the initial snapshot and move into his low cold six, there to half-roll into him – this generally resolves into a rolling scissors, going up, with an interesting near-stall situation on the top. Whoever burns the most energy in this part of the fight is likely to flop down first with the opponent hard on his tail. In the case that he makes a gentle zoom away, I may either do the same to increase separation and match his zoom foot for foot, or haul after him if I've already earned enough angles on him, for a stern chase.

In a long range encounter at max speed it doesn't make much sense to stick to a level merge. In such a situation it's far better to climb at a good pace with sufficient lateral separation to ensure that the upcoming merge is conducted at corner velocity, should he match the climb. If the opponent doesn't climb with you, you'll have an altitude advantage but will be at a slightly lower speed (given a co-E starting position and similarly performing fighters). In that situation the bandit generally goes for a pipper-on zoom which you can defeat by spiralling around his turn vector near stall speed (see Chapter 13.5, The Rope-a-Dope). The margins are extremely small however, so you really need to pay attention to his energy state and attitude in defeating his last-gasp shot. What follows after that is either an immediate minimum-range engagement, going down, or, if you're careful, a continued spiral climb to establish a significant energy advantage while the bandit flops down at the top of his zoom.

There are any number of "ideal" tactics to employ in a classic co-energy head-on merge between similarly capable fighters. The problem is that those situations where you can employ them are few and far between. Usually you will face odd and disjointed situations with the enemy coming in before you're prepared, where you're forced to manoeuvre with what you've got in terms of energy and possibly a damaged crate, where there are more than one bandit in the merge to reckon with and so on. Thus, approach each situation with a fresh mind and determine the particulars as they are given: who has the energy trump, what are your options, what are his options, does your aircraft dictate a specific tactic to be employed, do you really want this fight, where is your exit etc. There are no fast and simple answers – it's all in the situation as it presents itself and how it unfolds. One thing I can say, however, is that it is exceedingly important to establish an immediate and powerful psychological advantage early in the fight. If you can scare or rattle your opponent, if you can flaunt your absolute confidence in yourself, if you fly effortlessly and laugh scornfully at the pathetic flapping of your enemy, then you will have conquered him already. The actual shooting down business is merely something that happens as a matter of course.

How do you deal with a one-versus-two situation where the bandits are somewhat separated, i.e. coming at you with some distance between them? Do you turn with the first or with the second? Do you turn with the second guy, or do you refrain from turning altogether? Well, assuming that you're co-E and equally capable (rare, but it happens!), fighting the first guy is a definite no-no, because, if you reverse you're setting yourself up for an easy shot by the second guy. You will want to pass the leader as closely as possible, inside his turn radius but defensively enough to avoid the collision and the head-on shot, essentially forcing him to pay a steep energy price when he reverses after you. Just think "fly past, no guns" to avoid the collision. This will limit your options against dude number two, since he's probably flying pipper-on-you. Thus he too is coming at you more or less head-on. Meanwhile guy #1 is busy reversing. I normally blow by #2 as well and keep going, for should you zoom on dude #2 you will most likely get nailed near the top of your zoom, or at least patently disturbed, by the leader. That fellow too (pesky guy #2) will reverse expensively, leaving you with a slight energy advantage over both of them. Maximise it by cleaning up (unloading the stick) and if you do have the extra edge, which is plain to see after a few seconds of fruitless tailchase on part of the enemy, consider whether to chandelle them or to make a clean level getaway. They're out of range, so no worries, and you have plenty of time to radio in help.

If you're feeling gutsy, by all means rope them in. However, you will need quite an energy overhead to be safe against a plain old cut-across-the-circle zoom: when you chandelle around them they must be unable to close to guns range at manoeuvre speed. Since you may be unable to secure that much of an advantage following the first neutral merge, you will have to fake them into breaking or stalling (by executing a fake rope-a-dope or by making a very aggressive high-sides run after your chandelle) so that you can make another clean extension before setting them up anew. And so it goes.

17.2 THE BOUNCE

Setting up a bounce is always a pleasure. How to execute it is highly dependent on the overall situation however, what type of quarry you have before you, relative energy states and flight profiles. If the target is a fast mover you may have to drop in from whatever angle is presented and switch between lead and lag pursuit to line him up properly, and the attack is generally quite shallow and predictable. If the target is slow you have the luxury to manoeuvre around to his stern and set up a proper run. In any case, you generally have the option of choosing between a high six, a level six and a low six bounce. They all have their respective pros and cons, so which ever you choose is really up to you and the situation as it presents itself. In these examples we will assume that you have an unholy amount of smash to begin with and that the bandits are totally unaware of your presence. In most such bounces you will have to reduce throttle in the initial dive so as to avoid compressibility effects and potential damage from disregarding VnE[1].

THE HIGH SIX BOUNCE.

The high six bounce is delivered at an acute vertical angle to yield a shot which offers as much enemy planform as possible, i.e. a big fat target with lots of exposed wing and the entire fuselage area. You will be coming down at anything from 60 to 30 degrees angle off tail in a near-straight lead pursuit approach where your elevator input is minimised to concentrate fire in an unloaded state. Since your closure is great you will need to open fire at about 3-400 meters and press the attack to minimum range in order to deliver a crippling blow. Depending on the target's vulnerability, i.e. location of engine(s) and fuel tanks, you should aim either for the engine and cockpit area or the wingroot, taking care to draw proper lead according to target rate and distance. You're coming in fast, so aim just ahead of the propeller and let the target slide through your gunsight. As you close to minimum range, break off the pass by letting the target's own speed bring him up through your gunsight. Thus you slide

[1] Velocity Never Exceed, i.e. the airspeed at which you will lose control through compressibility and/or suffer catastrophic damage such as a wing torn off.

behind and below him momentarily before bottoming out of the dive and zooming away, either to disengage or to re-engage for a subsequent pass.

The high bounce is difficult on the beginner inasmuch closure rate is generally high, leading to a shot opportunity of brief duration at high angle off. The canny defender can rather easily move out of the guns envelope with a deft roll and jink, or even a deliberate skid, provided of course that he sees the attack coming in due time. Moreover the beginner all too often falls prey to target fixation in the final moments of the bounce, causing a rather painful collision instead of a clean kill. If you sense the shot going awry, resist the urge to correct and let go instead to set up a fresh run. Furthermore, be advised that if you're coming in with a massive energy advantage, it is auspicious to chop throttle in the dive so as to avoid compressibility effects, to further increase your shot window and to avoid bottoming out far too low of the target.

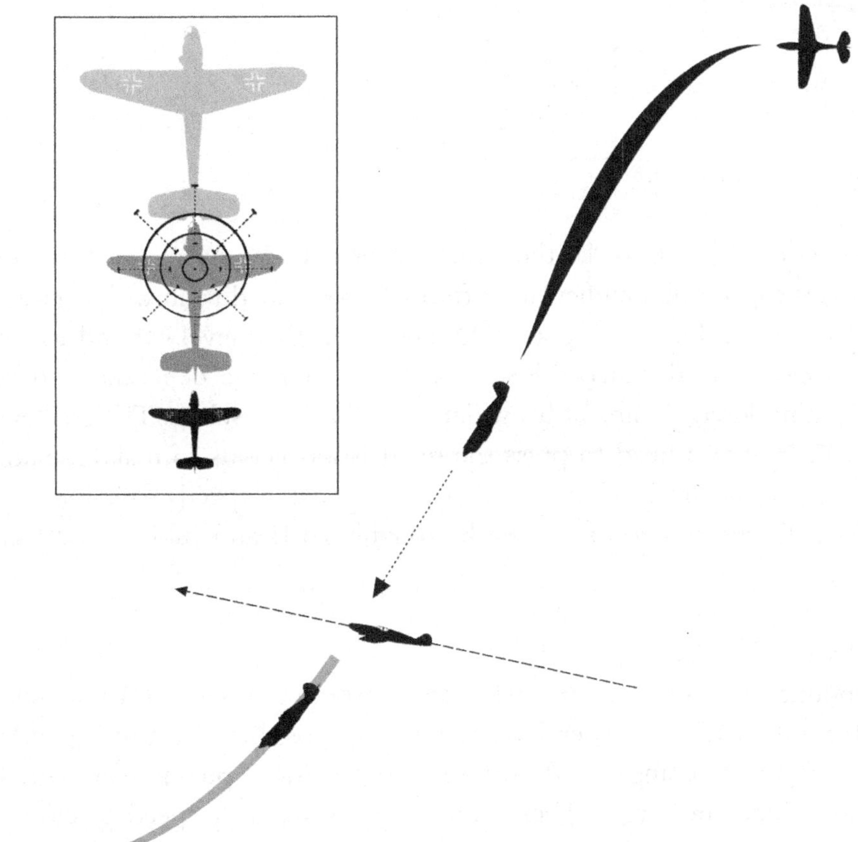

Fig 50. High bounce (not to scale). The gunsight sequence shows how the gunsight remains steady at an unloaded state, allowing the target to motor into and through the shot window at decreasing range.

THE LEVEL BOUNCE.

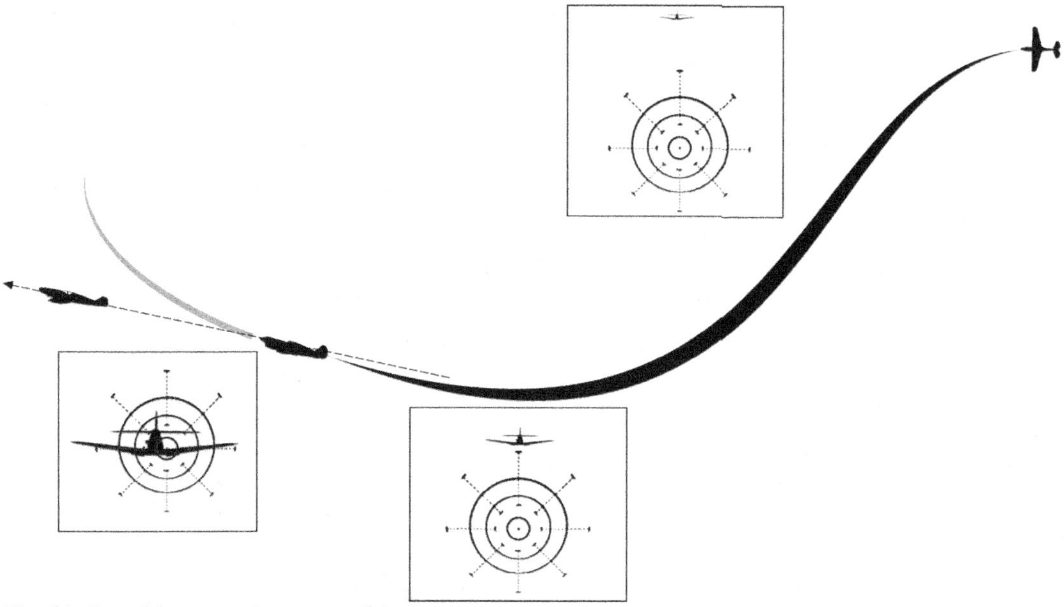

Fig 51. Level bounce (not to scale).

To make a level bounce you must let the target motor out a bit. When he's about 45-60 degrees off vertical and perpendicularly offset to your side, dive in with lag pursuit until you're about to hit the enemy plane of manoeuvre, then level off and use your excess speed to close the distance. Close to convenient range depending on your closure and gun him down, aiming at his wingroot or centre fuselage. The greater the closure the farther out you need to press the tit. It is auspicious to make the attack with good closure and to break off high, rather than to match speeds for a long firing window, because if you miss you will then be roughly co-E and subject to effective counter-measures.

THE LOW SIX BOUNCE.

The low six bounce is the most profitable as it exposes you the least to enemy observation. Here the trick is to spend as little time as possible in his high and level six by diving in at an acute angle, and bottoming out without pulling too many G's well below and behind the target. There you use your superior speed to close the distance in his very low and very cold six, the place which most pilots have a hard time to clear visually. Manage your closure with throttle as necessary so that you get a comfortable firing window. The shot will be delivered at a greater angle off than in the level six bounce, allowing you to strike harder at the softest part of the quarry. Break off high and observe.

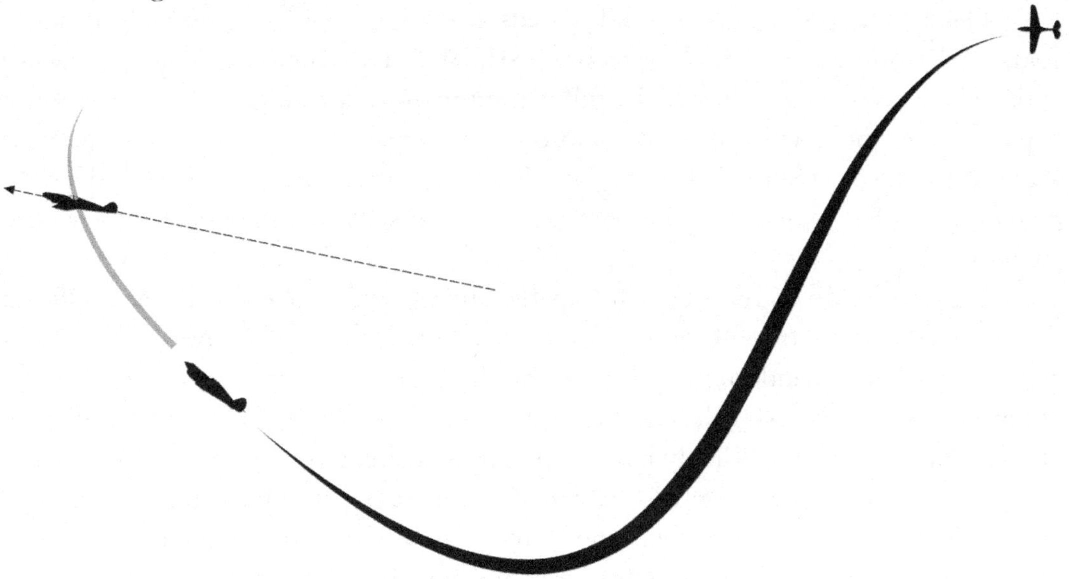

Fig 52. Low six bounce (not to scale).

So, tooling in at 17K near the enemy base, I spot a sleeper down below, happily climbing to whack Yanks down south, or so he thinks. Boom time! I wing over and go in, throttle fully closed, and dive way below in his low six. It's a big black brute. He appears to slip somewhat, checking his six, but I'm fairly certain that I'm well below still. Closer... closer... it's an Aichi D3A2 Val. When he's fully in my sight, I tap the cannon and let fly two 37mm rounds in quick succession. The first narrowly misses, but the second hits him squarely at the starboard wing root. Whoof! A puff of grey smoke! Surely he must blow? Aye, as I swing clear, a solid black burst flit past below my port wing as -toad- falls messily into the sea. Poketa poketa! Muahahah!

WarBirds, December 6 1998

WHAT DO YOU DO IF YOUR ATTACK IS DETECTED?

You have spotted a con and manoeuvred to put yourself into position for a high six bounce. You commit to the attack and is just about to close in to guns range when the enemy is seen to react to the immediate threat. It's usually readily apparent: a slight dip of the wing, a minor yaw, a sudden shudder, a significant course correction or attitude change, tell-tale signs of the enemy getting wind of your attack and preparing to meet it. Most telling of all is of course the radical guns defense break, usually executed high and toward you. If you're coming in from dead six or only slightly offset from high six the enemy may break in either direction, high or low, left or right. If you on the other hand are closing from low seven or level eight o'clock, the bandit is likely to break high to port in order to maintain visual contact and put his lift vector on you. It goes against the grain to do anything else, that much you can count on.

Depending on the range and speeds involved you have two fundamental options: to continue the attack or to break off. If you continue the attack, you must choose between a single blow and a full commitment to a turning fight. This again is dependent on the aircraft matchup, and on the energies and vectors involved. Your actions are also governed by the amplitude of the enemy reaction: does he manoeuvre radically or gently, at what rate does he start to move out of your guns solution?

Let's say that you're in equally capable aircraft and that you have a significant energy overhead – after all, you're coming in hot. If the bandit breaks hard it is generally best to abandon the attack right there and conserve your energy for a second go rather than to seek a fleeting snapshot. If you do fall for the temptation to line him up you will most likely have to pull immediate and increasing gravity loads, causing your shot to arc heavily behind the bandit as he zips laterally away and around. This robs you of precious energy and may offer the bandit a chance to reply with a rolling scissor as you overshoot behind him. If the bandit breaks late he will most probably not have had time to prepare a rolling scissor: he will most likely content himself with getting out of your guns, then to recover and plot his next move. In any case you've missed your shot and wasted energy that could have been put to better use in an extension or vertical separation. If on the other hand he sees you coming from a ways out and delays his break until the most profitable moment, you can be pretty sure that he's planning an immediate reply rather than a panicky break: act accordingly!

If you're savvy enough to break off the clearly unproductive attack and subject the enemy to repeated assaults, you may eventually notice that the bandit employs the same guns defense manoeuvre against all your attacks. Use that knowledge to your benefit: you can fake an attack and break it off early to improve your energy advantage at the very instant you see the bandit go into evasives, or you can plot a better attack run that is aimed at the point where the bandit shows his planform rather than to advertise a standard pipper-on attack against the still holding enemy. Then again the bandit may do the unexpected and break away in a diametrically opposite direction, causing you some discomfort. If that is indeed what he does you can expect a quality fight against a cunning foe rather than a simple slaughter.

Personally, I usually restrain myself from pulling more than 15-20 degrees of deflection from my original path if the bandit breaks, in order to retain energy. When I see that the bandit is producing a good rate out of my solution, I pull only so much as to overshoot the bandit as closely as possible, at a high angle off relative to his track, so as to defeat a likely riposte (the smaller the separation the less time he will have to settle down in the gunsight). As I blow past him I know that he will have spent a good chunk of energy in his guns defense, whereas I have saved the bulk of my smash for a follow-up.

If, on the other hand, the bandit only produces a low rate of angular displacement out of my guns run, within the 15-20 degree parameter mentioned above, then I'm free to adjust to lead pursuit and unload (release stick pressure) for a small angle-off snapshot. The trick here is to gauge his intent and judge his future flight path, and to correct the course so as to intersect the anticipated path with a solid burst of unloaded fire. A quick jink or a deft roll followed by a short pull on the elevators to line up steadily towards the intersection point is what makes my day and ruins his.

Now, let's say the bandit is on the ball and starts his break early, well before you close to guns range. The absolutely worst choice is to keep at it and draw lead as this will generally degenerate into an equal opportunity head-on contest. In this case it is far better to opt for a high lag pursuit and plan for either a chandelle or an oblique reversal. E.g. you're coming in hot on his seven o'clock: he curves around obliquely to port (left) and shows you a full planform, obviously intending to swing fully into you and meet you head-on. You know that he's spent much energy in the reversal already, and that he'll spend even more in the next 180 degree reversal as you bore past him. In that regard you've already established a further greater energy disparity to your benefit. Instead of pointing straight at him, point your crate above and to the right of him and blast by at a massive clip as he commits to the second 180 degree turn. Now use all your smash to zoom away at a 30-45 degree angle, refraining from the temptation to bank and pull enough to keep him in view until you've passed squarely out of guns range. You already know that he's behind and below, and unable to follow. Maximise your zoom up to corner velocity. Once sufficiently separated, chandelle around him (probably to left since you left him in your low seven o'clock) and establish his whereabouts and intent. More than likely he will be trucking after you, well below and well behind. Now's the time to goad him into the rope-a-dope (see chapter 13.5) or to subject him to a series of diving attacks.

Most pilots err in persuading themselves that they absolutely must kill the enemy on the first run, as if this particular bandit is the only one around, as if only this particular opportunity is available. Hence they go "all in" at the first opportunity, whether opportune or not. Try to keep a cool head and use only so much energy as you can afford. You must realise that if any one particular card is seen and met with equal force, you must abandon the wager and move about to renew your attack at better odds. Remember that patience is at a premium. Rashness and over-eagerness will get you killed in a hurry.

17.3 SIX O'CLOCK HIGH!

Being assailed from six o'clock is most pilots' worst case scenario, quite needlessly. Assuming that you spot the bandit in time to manoeuvre there is precious little to worry about. Fact is, the high six attack is extremely easy to defend against, as we have already discussed to some extent in previous chapters. Now, as the bandit comes streaking down, the first thing to do is to NOT PANIC. If he's close to guns range already the decision is easy and reflexive: immediate guns defence! Break decisively toward him, or, in the worst case, 90 degrees and down, to defeat his shot. The idea is to increase your lateral (sideways) rate through his gunsight. By adding a slight diving angle to the guns computation, you make his shot extremely difficult. What's more, the time required for him to see, appreciate and compensate for your new vector is usually more than he has at hand when coming in hot from high six, resulting automagically in an overshoot. I.e. he blows by your wingline without a shot, placing himself in front of you (although you may not have a shot yourself). You KNOW this. Don't fear the bandit behind you. Laugh at him and treat him as your victim, not your assailant!

If he's still a ways out you have plenty of time to consider your gambit. My personal favourite is to play sitting duck – naturally while keeping him in sight – and to break for a barrel roll just prior to guns range. If that is out of the question because you've already demonstrated that you've seen him (as in making any kind of violent manoeuvre or in making a significant change of heading), then he'll quite naturally assume that he's been detected and can expect a break in your current bank, and thus you have lost the element of surprise already. If so, you might as well start your response without further delay. What you do depends on your plane, proficiency and mission: you might set up a scissor, a rolling scissor, a standard break turn or even a simple split-S to get out of Dodge, should the situation warrant it. Personally I ascribe to the notion that it's better to have a bandit on your tail than none at all, and prefer to duke it out rather than turn tail – but that's just me. The thing to remember is to turn the bandit's speed and vector against him: if he's well out to begin with, there's nothing stopping you from turning into him and forcing him to deal with a much steeper angle, or a perpendicular angle that he cannot pursue as that would require him to pull excessive G loads. Most likely he'll abandon his pass at that point, unless he's stupid enough to press it, and regain position for his subsequent attack. What happens after that is entirely up to the situation at hand.

If the bandit does press his attack, reel him in and defeat his shot, then immediately counter-attack by rolling back into him. Beware the risk of overdoing it, however. You will want to come out nearly parallel to his flight path with a suitable amount of lead for a long tracking shot. When you, "the hapless victim", act confidently to ruin his fine attack and even have the audacity to fight back, his morale will suffer a serious blow, even if you don't hit him, even if you don't fire. The mere fact that you can easily defeat his best attack and turn it into an opportunity for yourself immediately destroys any hope he might have had of an easy victory. His false sense of security switches instantly to a very real sense of fear and inferiority, it turns his bones to jelly and before long he will be duck soup.

The thing to watch for is the early break. That is, while you perform your violent or otherwise SA-taxing guns defence, the enemy breaks off his attack and repositions himself for another go. If you lose the bandit in this instance it is highly likely to go noticed, and you will then be forced to defend blindly against an unseen attack. That is very, very, dangerous.

If you do lose sight of the bandit, manoeuvre *immediately* to regain visual contact. If he's still some way out, a quick quarter-roll and pull on the elevators suffices to unveil your hitherto blind astern. If he's close by, corkscrew and observe your high and low stern quarter. If he's not there, he'll be somewhere high to the side or directly above. If he's not there, he's high in front. It only takes a few seconds to catch sight of him again – and if you don't, continue the search by barrel-rolling until you're positive that the badguy is nowhere around. Know that it does happen from time to time that the bandit abandons the attack altogether or is otherwise lost on his own accord.

17.4 THE EXTENDING BANDIT

You're chasing a bandit whose energy and/or performance is substantial enough to level or take away your closure, still you persist in the vain hope of him turning around to give fight. Ain't gonna happen, at least not until he's dragged you through sheets of flak or until he's increased his separation enough to dare turn back. Yet you press on, hoping beyond hope or fearful that he'll turn around and pluck you from the sky the moment you let him go. Right there is your opportunity! If your closure is nil or negative, make a highly conspicuous break away – he's sure to be looking – as if you were giving up the chase, all the while keeping him in sight. Make your turn about 30-45 degrees, quickly roll level or nearly so to keep him in sight, and throw in a vertical element too for shits and giggles. Of course you have no intention of letting him go, you're only giving him that impression. The moment you turn away you can be reasonably sure that he'll cackle with delight and come after you. This is precisely what you want. If his energy is superior, give him enough angles and closure to entice him deep into the chase, then reverse into him. Or give him even more rope to hang himself with, by presenting what appears to be a simple dead six shot – you're playing sitting duck only until he draws just outside guns range – then bite into him. Most of the time he'll take the bait and turnfight you, which you're aware of and already converting your first break to a lead turn. If his energy isn't all that superior from the get go, you might as well reverse your false break away turn immediately and go for a frontal lead turn as previously described, and then directly to the endgame. Piece of cake!

17.5 THE LOW QUARTER LAG

Sometimes you just don't have the smash to catch a bandit, but the difference in energy is not great enough to give him an immediate advantage. The situation is generally such that you're stuck some 1000 yards behind while the extending bandit is in a gentle bank to keep you in visual contact, and he's most likely climbing a bit as well. Rather than staying in pure pursuit at that point, which is an exceedingly unproductive thing to do, or risking a potential rope-a-dope by cutting across his circle in lead pursuit, you might want to go to lag pursuit instead (i.e. point your crate below him and opposite to his bank). This puts you in his blind spot below the tailplane and forces him to spend a bit of energy to reacquire you, and if he's really inept he might even burn a whole lot of smash by breaking hard in any direction. In such a case it's fairly simple to cut across the intervening distance and convert your energy to position with a continued lag chase and subsequent vertical reversal. Just take care to avoid a potential head-on situation and you'll do just fine. If, on the other hand, the bandit isn't deceived by your cold lag repositioning but hangs on to his energy and starts a proper rope-a-dope, well, then you'll just have to deal with it as previously discussed (see chapter 13.5).

17.6 THE VULTURE

Sometimes you're out of luck and must launch from a field where vultures hang about to smack the unwary. A typical rookie mistake is to accept that situation at face value and start fighting soon as you're wheels up. Bad mistake. When you're just airborne you have absolutely no smash to manoeuvre profitably. The proper thing to do is to sneak out any which way you can, preferably just as the vulture concludes a pass one some other hapless bastard and must spend some time to regain altitude, and exit the general area in the weeds. If he comes after you, fine, go for an energy-efficient guns defence. If he doesn't come after you, even better. Keep on going and secure altitude sufficient to deal with the vulture on equal terms or better, then return to sort him out with a good head of steam. It's eminently easy!

17.7 GETTING STUCK

Getting IN a fight is easy. Getting OUT of the fight isn't. What if you, despite having wowed never to get stuck in a furball, get stuck in a furball and cannot deal with it? Wheeling and cavorting in an almighty mix-up of friendly and enemy fighters can be a lot of fun to be sure, but it's seldom conducive to longevity. Furballing is arguably

THE hardest thing you can engage in, and the most risky not only due to the immense task overload it produces but because of the extremely high risk of collisions, and not least the likelihood of catching an unlucky snapshot. Thus you will want to disengage as soon as humanly possible. Fact is, you will want to disengage the moment you engage.

First of all you must cast aside every thought of hurting the enemy and concentrate on rolling and scissoring your way to the fringes of the furball, and then to keep barrelling your way out altogether. In some cases you may disengage furtively without drawing attention, whereupon you may straighten out somewhat and add to the distance. In other cases you may acquire a "groupie" or four who sees an easy kill trying to slip away. In such a case you have no option but to keep opening up separation from the melee while dodging your pursuers' attention as best as you can – barrel rolls are great for this – then to deal with the remainder as soon as you're out of sight from the melee.

Another great way of getting out of a general melee area is to dive out, right through the brouhaha in an aileron roll, then to scoot out in the most unpredictable direction. You're highly likely to attract a couple of groupies anyway, but here you have at least got some energy to work with as opposed to trying to slip out furtively. Remember now, staying in or in the proximity of the furball is NOT an option. Running straight with multiple enemy in guns range is NOT an option. And running while attempting some sorry excuse of distance-covering mild evasives is NOT an option. You have to fight, and fight on your terms.

Dealing with three or four pursuers isn't very much different from dealing with a single bandit on your six, you'll just have to work a bit harder is all. The good thing about it is that they're all on your six – that much is certain, and that is a lot better than having them all around you. The bad thing is that they're all firing at you and getting closer all the time. Well, say the fun that lasts!

Assuming that you're set upon by a gaggle of bandits here, you must try to reduce their numbers yet further. Rolling scissors is great for messing up peoples' SA, so use that to throw the SA-challenged in the pursuing party whenever the competition gets too close for comfort. You'll soon see the SA-deficient wander off without a clue as to where you went! Furthermore, those of your pursuers who fail to stay with you in the scissors will have to spend precious energy in recovering and reacquiring you. Exit with the remainder using the time-honoured barrel roll until you're down to one or two bandits, then start picking them off. Note though that you should expect to make a lot of wholly or partly unsuccessful disengagement attempts before being able to reduce the opposition to a manageable level. Because, while you're busy trying to dodge the last remaining pursuers, those that you threw off initially may have gotten the scent again and rejoined the chase. If you have some distance (and terrain) to play with, make an abrupt heading change to confuse the distant pursuers yet more.

While running away you have to take stock of the situation: how many enemy, how close, what types, what options? Keep your cool! You've thrown the gaggle and you've thrown the bulk of pursuers already, dealing with the last two or three shouldn't be too hard. OK, three is cutting it closely but two isn't much of a challenge! Flummox one of them by switching from straight out scampering to engaged manoeuvring – odds are you'll catch one of them with his pants down and unable to convert from easy pursuit to hot combat. Watch him sail sightlessly away or hurl himself disorientedly into the ground, then exit with the remaining two. Once they start getting into range again, jink and yank slightly to put some more distance between you and the thrown party, then convert again from supposedly panicked flight to fangs-out fighting.

My personal favourite for converting from flight to fight is the rolling scissor closely followed by the oblique spiral, either of which are extremely tough on the pursuer. The oblique spiral or corkscrew requires a crate that turns well, however, preferably better than your opponents', and one that doesn't stall too easily. If you're in a plane with high wingloading you probably won't be able to pull the requisite AOA or spiral tightly enough – go for the rolling scissor instead. The scissor usually leaves you fighting one guy while the other tries to gather his wits, enough for you to send the one packing or burning while preparing to receive the other. If you do manage to clean out one of them you can be Pretty Damn Sure that the remaining guy will be fairly stressed out and ripe for plucking.

When corkscrewing you're presenting a very difficult target, but you're also choosing to become more or less stationary. Thus you give the bandits an opportunity to zoom and subject you to one boom and zoom attack after another until you're down and dead – on the other hand, few bandits can resist the urge to plough right in and gun you down in the spiral. If they opt for the latter, you'll be spiralling around their performance vector – actually execute a rolling scissor in the vertical. Use the spiral against the leader just as he comes into guns range but be prepared to switch over to the trailing guy – he's your priority target at this time as he has more time to react to your move and will arrive in guns range when you are at your most vulnerable. The spiral must be extremely aggressive, you have to develop max turn rate and max climb rate while keeping your wingtip pointed at the enemy, and utilize every last ounce of energy to twist onto the back of the vainly pulling bandit.

Some of my most memorable fights have been disengagement battles, singly or alone. I remember particularly one funny episode where I, in a slow but well-turning Curtiss H75's (P-36), fought a 109E somewhere deep inside enemy territory. My wingman having augered in or been shot down and myself completely out of ammunition, I had to break free and return to base. Only the 109 wouldn't let me, and being 50 km/h faster he had the means to enforce his decision! So we would turn a couple of flat circles a few feet off the deck until I could climb onto his back and show him the futility of turning with me, he'd try to shake me off without success, and finally he would try to make a run for it, at which I demonstrably turned away for home. And of course he would double back to try and shoot me down. I would let him close until just outside range, then swiftly spiral onto his back and remain there until he scampered off again. I believe I had to demonstrate this a good six or eight times before he finally lost sight of me in a reversal and I could make good my escape. Well, he got A for effort at any rate!

If you succeed in breaking free from a 10+ bandit melee, and as long as your crate isn't in too bad a shape or extremely underperforming that isn't too hard unless of course the bandits are Sierra Hotel[1], you can safely regard yourself as having attained a very high level of SA and proficiency. A final tip: when fighting a gaggle or several opponents, make sure to kill the best guy first. Leave the sad sacks for last – they aren't much of a threat anyway!

17.8 COMMON ERRORS

Every sortie is a learning experience. Hardly any sortie is 100% faultless but once in a blue moon you'll come home with a bagful of kills and make a perfect landing, knowing that you did everything right. And then again, you probably didn't. You almost collided with one guy, you wasted your ammo at one point, you lost sight, you failed to capitalize on an advantage, you lost your wingman und so weiter. There's always room for improvement, and taking stock of your performance both during and after the sortie is material to your gunfighting progress. Get used to writing after action reports, use gun camera programs to analyse your gunnery, and keep track of what sort of error caused this or that loss of life. If you don't, you're destined to be stuck in a rut and will eventually become bored and/or disgusted with yourself and/or the game.

WISHFUL THINKING

The main fault of nearly all casualties is to accept a bad fight despite better knowledge (or due to lack of the same), or that of sustaining a manoeuvre that has already been proven faulty. Some pilots won't even break off when they're taking fire – incomprehensible to me – because they still think, or worse, hope, that they will somehow magically prevail. Hope has no place in air combat. Once you see that you're being out-turned, once you see that odds are piling up against you, you MUST

[1] Shit Hot

take immediate and radical action. Don't wait for the final confirmation, for it is highly likely to be accompanied by a fistful of shells. The sky is full of inept pilots who fail to recognise when a situation is about to turn sour – you will see them all over the place, flying predictably and benignly, not checking their six or worse, disregarding obvious threats. Gun them down without compunction. Another case of wishful thinking is when you're being chased off a scene by a faster aircraft: the great majority of pilots hold off manoeuvring until the enemy is in guns range, whereas it would have been infinitely more productive to start manoeuvring while they still had some separation to play with. Likewise, once a pilot loses sight of the enemy he is apt to STOP manoeuvring – "out of sight, out of mind" seems to be the case here – whereas the correct course of action is to manoeuvre radically to regain visual contact without delay. A steep spiral looking around and behind is usually enough, although you don't want to do that in a particularly bandit-infested area. In such a case, a long barrel roll or a couple of significant turns should suffice to clear you visually and regain visual contact.

FIGHTING THE ENEMY'S FIGHT

It's easy, particularly when aircraft performance is dramatically different (such as a slowpoke Hurricane against a FW-190 or a zippy Thunderbolt against a Zero), to lapse into fighting on the enemy's terms. It really doesn't matter whether you have the advantage or disadvantage in any performance area, or whether you're closely matched: never ever accept a fight where the enemy is allowed to use his best trait against your worst. To the contrary, you must impose such a condition on him. If you're in a fast, high wingloaded aircraft, you must use energy tactics against an angles fighter, and conversely if the situation is reversed. Similarly, you must learn to identify the enemy intent – such as seeing an upcoming rope-a-dope or a pending drag-and-bag situation – and move to avoid it. This all ties into knowing your own and the enemy crate inside out, of correctly assessing energy states, and correctly identifying the opposing pilot's proficiency and game plan. Once you accept a fight on the enemy's terms, you're becoming predictable and an accident waiting to happen. Break out of it and manoeuvre so as to gain the initiative, or disengage entirely.

HEAD ON

You will already have noted that rookie pilots tend to fly "pipper on the enemy" until they learn otherwise. The most common, and most disappointing, example of this is the pure head-on merge which plays itself out again and again with equally predictable results. You must realise that the head-on attack is the worst imaginable kind of attack. It's Russian Roulette at best. If you fly a steady path right at the enemy, and he does likewise, the odds for your survival are as low as his. Even if you don't take damage the risk of collision is considerable. And even if you happen to outgun the enemy by a factor of 10:1 the odd chance of a single bullet entering your brain simply isn't worth it, and it does happen all the time. The thing to do is to bank on the fact that the bad guy will make this rookie mistake, and use it against him with a lead turn that defeats his shot and puts you on his six o'clock before he has a chance to recover. Never ever, EVER, accept a head-on encounter – not in the merge, and never in a nose-to-nose turning fight. Fly to defeat his head-on by putting your velocity vector outside his flight and guns envelope, and you will never have to suffer either a guns death or a collision. I cannot repeat this often enough: never accept a head on!

FIRING TOO SOON, FIRING OUT OF RANGE

It's self-evident really that you should only fire your guns when you have a reasonable chance of hitting the enemy. Anything else is a waste of ammunition and, what's worse, tells the enemy volumes about your lack of proficiency and of your uninspiring mental state. There's but one exception to this rule, and that is when you're closing fast on the target or when you have a fleeting snapshot. In both these cases you need to open fire shortly before maximum lethal range or shortly before you reach the point where you must open fire to converge with the anticipated path of the bandit, so as to maximise the volume of lead at the most auspicious moment. Firing too soon in a dead-six chase only tips off the enemy: get closer and closer still. Firing out of range, to annoy or in the hope of maybe hitting him with a bullet or two, is the mark of the inept – just don't do it.

HELPING THE ALREADY LOST

Throwing good money after bad makes no sense. You will often happen into the situation where a squadmate or otherwise friendly fighter does the inexplicable and bites off more than he can chew, usually at low altitude and at the worst possible place: in enemy transit corridors or over enemy-held localities. You may be able to help him for a little while by booming and zooming through the furball, but the results are usually poor at best. If you commit fully to clearing enemy from his turning crate you are more than likely to be wrestled down yourself – the odds are simply too bad. In that case it is far better to leave the poor friendly to his own devices and cut your losses. It may sound cynical, and it will make you feel bad for a while, but it's the only sane thing to do. In short: if YOU get in a fix, YOU are responsible for getting out of it too. Clearly there are borderline cases: sometimes two can prevail over five, sometimes just a single bandit is too much to handle. You make the decision, you live with the consequences.

RUNNING AWAY

Given notice of an assailant on his six, whether firing or not, normally throws the fledgling pilot into immediate panic and starts him into headlong flight. Away away! You have probably already experienced that turning your back to the threat and pretending it isn't there, or, equally bad, thinking that you will somehow magically outrun him simply by going to max speed, simply doesn't work against a bandit with an overhead of energy. Similarly, when the distant bogey draws closer, the fear-stricken rookie will point his nose down and scream earthwards in a power dive, until there's no more altitude left to spend. Then he will continue the straight run at treetop height, narrowly missing ridges and solitary trees. I'm here to tell you that running away doesn't solve the problem, and yields little knowledge too. Sometimes, most of the time actually, you will have to fight, or at least make a manoeuvre to facilitate flight – such as a narrow head-on pass or a rolling scissor which flushes the enemy sideways, allowing you a clean exit window.

For some reason, beginner pilots run away until there's no place left to run. E.g. while conscious of a closing enemy on his six, the defender continues a level and straight flight until the bandit draws within guns range. That's no way to do it! If you're clearly being run down like a dirty dog, start your manoeuvring while you still have some chance of recovering the initiative. A barrel roll attack (rolling scissor) works like a charm in these situations, as does the Corkscrew if you have the requisite manoeuvrability. Never wait to hear bullets and cannon shells slamming into your sad crate, if you are to go down you should go down fighting, not running like a wimp!

"JUST ONE MORE KILL, THEN I'LL RTB"

Greed kills. Need I say more?

PART V: NEVER ALONE!

A single fighter is as good as dead.

Flying alone is the norm in the online skies, for several reasons. First of all, players are undoubtedly individuals who sit around their lonely computers and decide for themselves when to launch and where to fly. Joining up is anathema to the great majority of sim pilots. Secondly, few sim pilots are even aware of the stunning benefits of teamwork mainly because they haven't studied air tactics at all or because they believe that the ornery procedure of finding someone to fly with isn't worth their while. After all, all they want is a few hours of good old fashioned entertainment. Thirdly, those who are aware of teamwork benefits and may even have experienced it a couple of times, perhaps as a target if nothing else, still prefer the lonewolfing practice because flying as a group requires discipline and a level of professionalism that more resembles work than leisurely entertainment.

To be sure, formation flying and formation combat does require discipline, restraint and a greater level of proficiency than going it alone, no mistake there. But once you've mastered the practice – and it's far easier than you think – once you've experienced the benefits and reaped the rich harvests of teamwork, you will never want to fly alone again, ever. Outside of that, flying as a group, in formation as it were, adds a whole new level to online combat that is simply too beautiful and enriching to pass up. The pre-flight routine of lining up on the runway, of assigning wingmen and going through mission particulars, the "start engines" command, the thundering roll down the runway and the soothing, almost hypnotic, sight of a squadron arrayed in combat spread, bobbing in time and weaving like a flying mat against the wind, is quite simply awe-inspiring. I wish everyone would have such an experience all the time. And there is truly nothing more exhilarating than engaging in a formation versus formation fight, regardless of the outcome.

The difference between formation tactics and single pilot tactics or even gaggle tactics is as that of night and day. Whereas the survivability of a single pilot depends on his machine and his ability to control it (and his intelligence), the survivability of a unit depends on the ability of ALL its pilots to work in concert. For this to happen, the unit must be trained and act according to a theory which your ordinary lonewolf generally finds far too taxing - it must be disciplined, it must act promptly and as a body with a single consciousness. Whereas the lonewolf engages with abandon and employs a wide variety of individual manoeuvres such as rope-a-dopes, hammerheads and what have you, such antics are simply not on when fighting as a unit – because the unit fights in an entirely different fashion.

CHAPTER 18: CROWDS AND GAGGLES VERSUS FORMATIONS

Simply attaching yourself to the nearest crowd (in the air) will not make you part of a formation nor will it substantially improve your chances of survival, mainly because such crowds generally doesn't have a recognisable mission or an acknowledged leader nor does it ever have an established internal structure. It rarely communicates and everything it does is predicated by group dynamics as opposed to authority or a singular will. You can hide in a crowd and draw some benefit from being but one of many available targets, but the overall efficiency of a crowd is small. It may dissolve at any time, leaving you high and dry.

Gaggles, on the other hand, are meta-cohesive units that mimic the behaviour of the organized formation. A gaggle may launch as a whole and have a semblance of a plan, though seldom one which survives first contact with the enemy. It may have a leader, but is seldom graced with a structure. Voice communication may be partial, loadouts may differ and deployment for combat is generally a great unknown. The gaggle usually becomes stationary at the first enemy encounter, and is seldom able to extricate itself at will. It is highly likely to fail in its mission, and few of its pilots can look forward to a safe landing. Needless to say, you should avoid being part of a gaggle if you cherish a long (virtual) life and career advancement.

Gaggles often end up flying a ragged line astern formation. This "string" formation is easy on the pilots as it doesn[1]t call for much expertise in maintaining, however, it also sets them up for swift annihilation. The Germans called the formation "idiotenreihe" - the idiot trail. And verily, it is just that. From the outset, the pilots are frantically attempting to close the distance to the flight leader and thus burn fuel at an alarming rate. Navigation is slow and cumbersome, and the flight leader is invariably barely even aware of the status of his force, because he cannot see them.

In the following story, I come upon one such gaggle:

Crossing the Zuider Zee and over the strip of land that separates it from the Dogger Banks, I spot one, two... six... seven, EIGHT specks ahead. Correcting west, I let down from 22 to 18K and pick up 300 knots. They inch slowly closer - they're level now, at 16K, and closure is GOOD. The old pump is thudding I can tell you! OK, calm down now, concentrate. One guy is straggling a bit to the left and rear of the formation, another is further ahead out on the left flank, there's one guy in sucked trail to the right... and a close gaggle of five ships in the middle. Coming closer, I identify them positively as 109-K's. Quick decision: who to take on first? The straggler. That will leave the others with 3-5 seconds to react, but I might get a snapshot at them as they break, or a full six salvo if I'm really lucky. Still no reaction. Are they all sleeping? Why don't they split? They MUST be sleeping, or concentrating on keeping that nice tight formation! Am I upsun? Perhaps! Sort of! I figure that the straggler has all of his concentration on joining up. Mmmm! I dip slightly lower and motor up... closer... closer.. 200 yards FIRE!

WarBirds, Jan 31 1999

The gaggle leader often rolls and weaves to scan the sky, which invariably breaks up the formation – because the guys in the back are trying to format at 2-400 yards (while those in the very rear are sweating to catch up). Thus, they are much too focussed on their leader and lose SA accordingly. When encountering the enemy, the pilots are often at a loss as to what to do and invariably become totally "locked up" or lost in a furball with the enemy – despite the best intentions. A gaggle usually fights without a plan – its only tactic is to bring overwhelming numbers to the fight and wrestle the enemy to death in a free-for-all melee where the formation as such is redundant – because every guy is more or less on his own. The outcome of such a fight is quite unpredictable: the formation may come out smelling of roses but it may also fall like so many flies. And enroute to the gaggle fight, the gaggle/line formation is meat on the table for anyone who happens to fall on them. A single smart bandit is all it takes to throw them into disorder and uncharted depths of low morale.

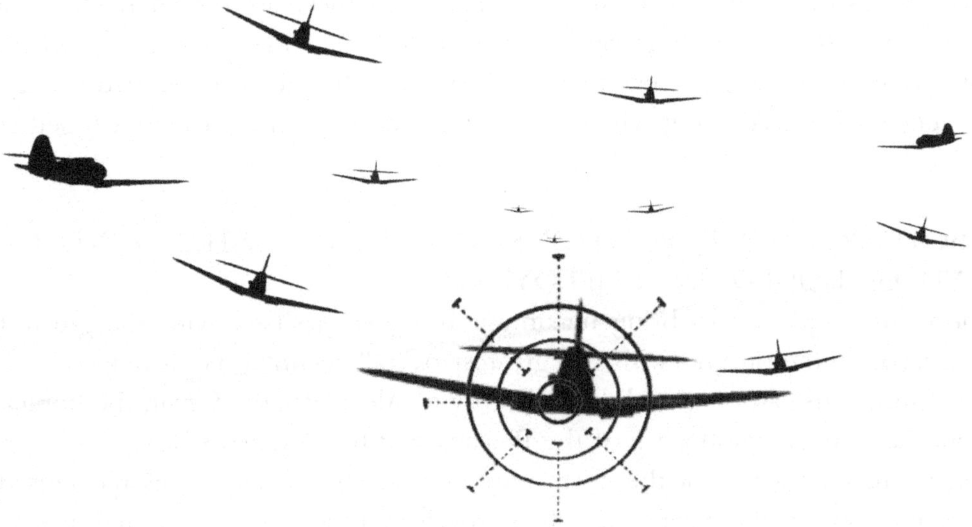

Fig 53. Coming in alone with a ton of energy on a gaggle, you can score at least three easy kills and get away with it – they're all looking forward, trying to catch up.

Pilot experience aside, single line astern or gaggle formations are extremely vulnerable because:

1. Despite its quantity of eyeballs, it cannot see well enough.
2. Despite its perceived strength, it cannot defend itself except by going static in a melee or by bugging out, both highly unpalatable options under the circumstances.
3. Despite its perceived cohesion, it cannot attack efficiently.

Formations then, are distinguished by having secured routines for all the key challenges so glaringly unaddressed in the cases above. It has a leader and an executive leader to assume command should the need arise. It has a strict internal organisation which clearly delineates responsibilities, tasks and deployment for combat and cruise. The formation launches as a unit with a uniform configuration and adheres to a strict formatting regime in order to guarantee its security. The formation engages deliberately and with a view to succeed in its mission, which measure is survival of its parts and as a whole. The formation communicates to a much higher degree than the crowd or the gaggle, and employs time-honoured formation tactics to accomplish its mission. That said, these are the hallmarks of a formation which does everything right. Far from every formation manages to do everything right all the time, or even some of the time. It takes a lot of effort and a lot of practice to succeed – but when everything comes together, it comes together beautifully.

WHY BOTHER WITH FORMATIONS AT ALL, AND WHAT KIND OF FORMATION SHOULD ONE EMPLOY?

Formations are employed to bring maximum firepower to bear with the greatest time/space concentration and with the greatest overall security. The origin of and cause for formations can be traced down history to Alexander the Great, the hoplite spearmen, the Roman legions' use of the Phalanx and the defensive Box. In the air, however, formations were not the norm until around 1917 when it was recognised that sheer numbers in the engagement had a decisive influence on the air war as a whole. From then on, formations have been an integral part of all military flying and it is indeed unthinkable to consider any situation where fighter aircraft will not be employed in at least wingpair formation. It should be borne in mind however that formations do tend to become a burden unto themselves if the numbers exceed that which can profitably be employed or comfortably directed. Since this is seldom the case in the online environment though, because online formations seldom grow greater than 4-8 aircraft other than in extremely well-managed scenarios, we can safely dispense with that discussion. One important distinction that remains to be made however, is the difference between a formation and a unit. A unit (which confusingly enough may also be dubbed a formation!) may consist of anything from 2 to 400+ pilots but doesn't automatically imply the use of any one formation. A formation in this regard is rather how the unit, ad hoc or not, deploys in the air.

Recognised units are more or less universally known as:
Wingpair or Element – 2 fighters
Flight or Division – 2 elements
Squadron (or Staffel if you're German) – 2 to 4 flights/divisions
Group or Wing (or Geschwader in German WWII parlance) – 2 to 4 squadrons

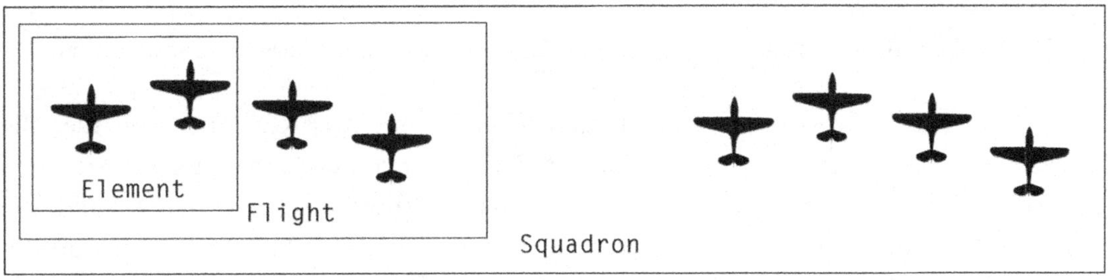

Fig 54. Unit denominations.

The most convenient formation for a 4-ship flight is the "combat spread", also known as "finger four" or the "schwarm" which is its origin. Maintaining a good spread requires a lot of practice, but once learned and mastered, there is no substitute. Other formations, such as Vee, Vic, Line Astern, Echelon or Double Column, are all decidedly inferior to the combat spread for various reasons mainly relating to command and control, visual security and tactical flexibility, therefore this author assumes that you will only want to learn the best practice and will without further ado dispense with covering the lesser alternatives in detail.

Before we go deeper into how the formation flies and fights, let's have a look at its smallest component, the Element.

18.1 THE FIGHTING ELEMENT

The Element is the smallest unit to take to the unfriendly skies and its fighting doctrine is also that of any greater formation – Flight, Squadron, Group. In the best of worlds you should never consider a sortie in anything less than an element and indeed that practice will become second nature to anyone who has ever seen the beauty, utility and efficacy of the fighting wingpair. Going out alone is simply not on.

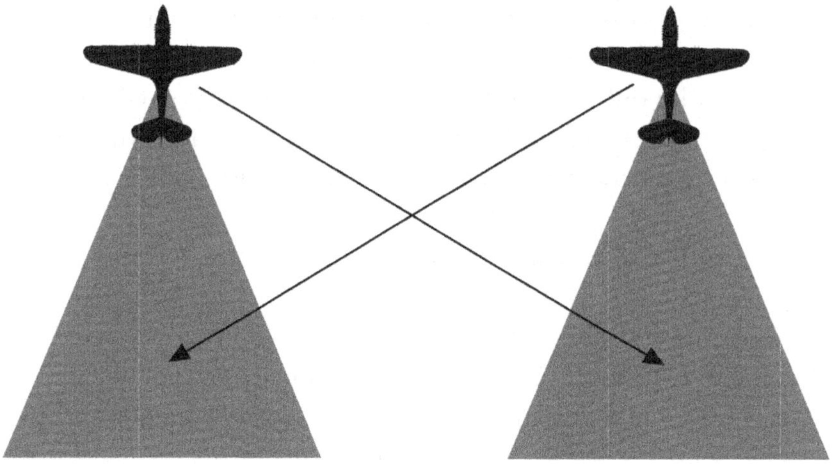

Fig 55. The benefits of line abreast formation. Cross-observation clears the blind sectors.

The standard formation for element is the combat spread, meaning that the aircraft fly beside each other in line abreast. The distance between ships is very much up to the situation and the numbers employed, although at no time should the distance be so close as to blind the pair from stern attacks. When cruising around in squadron or group formation and in relatively calm skies, the lateral separation can be quite small – 25, 50, 100 meters. When the formation is limited to element or flight, or when the squadron or group can expect battle at any time, separation should increase to 150-300 meters between ships. A squadron or group so disposed covers a whole lot of sky, and any formation in such a combat spread is virtually invulnerable to surprise attack from any quarter of the sky. This is due to the scanning procedure of the combat spread: rather than anxiously rolling and yawing around to keep the individual six o'clock visually clear, and spending an unholy amount of attention on formatting on a single leader up front, the formation flies coolly and calmly wings level in the given direction while looking over the wing to visually clear the neighbouring aircraft's stern quarter. This cross-scanning is continuous, almost to the point of the pilot looking exclusively over the wing rather than straight ahead. Certainly the pilot needs to scan all parts of the sky, including the forward quarter, and check his instruments too from time to time, but the majority of his scanning is over the wing and behind his wingman's stern. In such a manner every last piece of sky is always under observation, without the need for significant manoeuvres. Thus the blind spots below and behind the stern and below the wings are virtually nonexistent even for aircraft with notoriously bad six views. So disposed, the team can handle multiple bandits in any direction, at any energy state. However, while cruising, if you lose separation or position (i.e. sliding behind or drawing ahead) you're effectively blinding yourself and your wingman. Always work with throttle, separation and comms to maintain the spread.

Never ever fly in line astern formation. If you find yourself in trail, make a simultaneous 90 degree turn to regain combat spread, then return to the desired heading through a tactical turn, explained below.

TACTICAL TURN OR CROSSOVER TURN

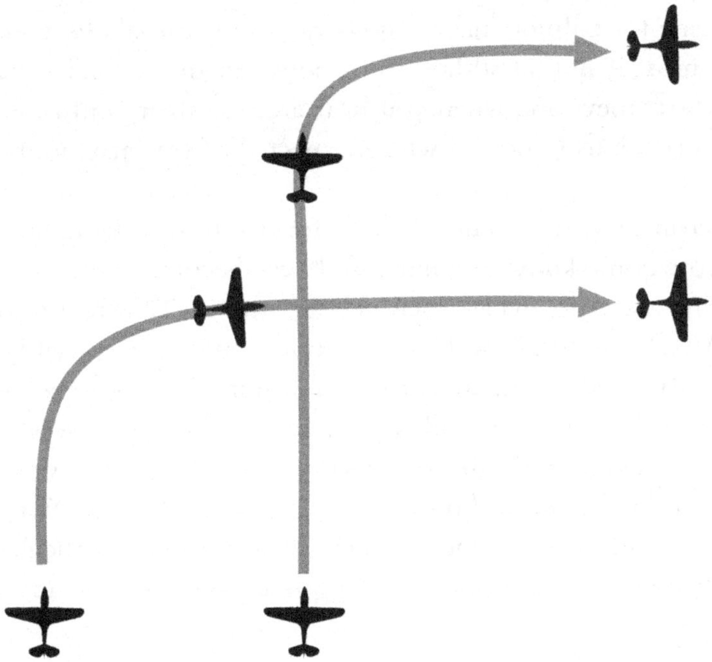

Fig 56. Tactical Turn (Crossover)

The tactical turn allows the element to change heading swiftly while retaining the combat spread and maintaining visual coverage to astern. The trick is to trade places in the turn by allowing the "outside" fighter to turn first. As the "inside" fighter, still on the old heading, observes his wingman sliding into his seven or five o'clock, he too initiates the turn. It's easy once you get the hang of it. Maintain speed throughout the turn - don't lose energy by pulling too hard. You may need to work a bit with throttle, lateral separation and small changes in altitude to reform. With practice, you should be able to turn together as swiftly and efficiently as you do on your own. The same practice is used for minor and greater turns than 90 degrees shown here. For a 45 degree turn, the inside ship holds his turn until he sees the outside ship pass behind and appear at about 20 degrees off his tail. For a greater turn, say 120 degrees, the inside ships initiates his turn well before the outside ship passes behind his tail in order to keep the spread after the turn. 180 degree turns are executed either as two

90 degree turns (which is rather cumbersome and time consuming) or as a cross-turn: the pair makes a simultaneous turn 180 degrees toward each other. Remember that turns should be made at a sustained rate, and level. While this practice appears laborious it has the distinct advantage of always keeping the entire sky under observation, even in the turn.

18.2 FIGHTING DOCTRINES

Having and following a doctrine for fighting may sound corny to the casual observer, but when you consider that most if not all ad-hoc formations out there don't have one and when you observe how they fare, when you hear and see their confusion, then you realise that having a previous notion of what's expected of you may not be such a bad thing after all.

If you don't have a doctrine, you don't know how to format, how to fight, how to support your leader – you don't know anything, and will become more of a liability in the fight than an asset. So, what doctrines are there? They're called "Welded Wing", "Double Attack" and "Loose Deuce" – but they might as well be called "Follow me!", "Stand off!" and "Separate and take advantage!". The separate doctrines all have their distinct advantages and applications, and have evolved naturally in step with experience and the opposition. While today most (consummate) pilots automatically fights the Loose Deuce fight, the Welded Wing and Double Attack doctrines still have some distinct advantages in particular situations. Thus, being familiar with them all and switching freely between them is a great advantage.

WELDED WING

In the dawn of organised fighter combat, in rickety biplanes over Flanders fields, and early on in the Second World War when combat experience was lacking both in individuals and in formations, wingmen flew close to their leaders and followed their every move as best they could. The general practice was to snuggle up with your wingtip but a few feet or even inches from your leader's, and somewhat below, and to stay there come hell or high water. Beyond sticking with his leader, it was the task of the wingman to look out for enemy and to keep his leader clear at all times, something which must have been an enormously difficult task given the slight separation. Entire squadrons flew such "welded wing" formations, twelve high-performance aircraft sweeping the sky as one single entity, normally in echelon formation. When combat occurred however, the squadron would shatter into wingpairs and sometimes not even that, because formatting so closely on a wildly bucking and manoeuvring leader took virtually every ounce of concentration and naturally pilots failed in keeping station. You can try it anytime you like in the online sky – I assure you it's no mean feat to stay formatted that way. It's a great formation to fly for airshows and simple ferry missions, but its application in combat is next to nil. There is one instance though when a tightly packed formation, particularly of the element and flight/division size, is useful and that is in attacking dense bomber formations and other static or slow moving targets such as troop concentrations and surface vessels. In such a situation the welded wing serves two purposes: it keeps the formation together for coordinated and repeated attacks, and it masses firepower.

DOUBLE ATTACK

Having experienced the shortcomings of the welded wing theory through practice, air tacticians understood that merely bringing numbers to a fight wasn't enough – you also had to be able to surround the enemy, attack him in sequence, and bring a ready reserve for contingencies. Thus evolved the Double Attack doctrine in which flights were treated as homogenous entities in their own right, and brought to bear sequentially to the fight. Thus, one or two flights would do the attacking business while one stood off to the side, ready to pounce or bail out the committed flights as necessary. Within flights and elements, individual pilots still flew welded wing. Still later, prompted by tactical innovation in the Luftwaffe, pilots of all nationalities learned to fly "Schwarm" or "Finger Four" formation within flights.

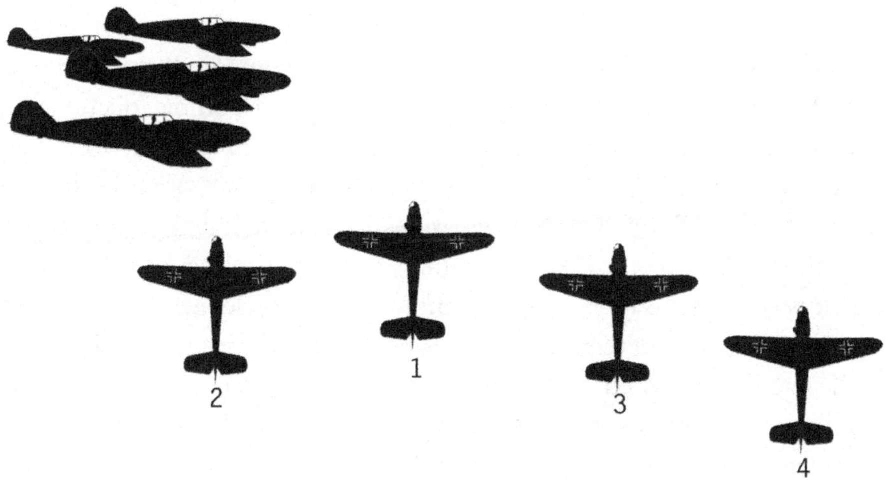

Fig 57. Finger Four (Schwarm). The formation can be flown at any distance between ships.

In the Double Attack doctrine, which also applies to the single Element's fight, one part of the formation (wingman, element, flight) "stands off" at altitude to watch against wildcards and ready to assist the engaged party when required, such as clearing out pursuing enemy when the engaged party disengages or recovers position. Such a tactic works fine when the combat is largely stationary and the combatants engaged in a typical wrestling match, or when the engaged party is subjecting a hapless prey to repeated slashing attacks. It is less suited for a co-altitude, co-energy fight, or for any fight which covers a lot of airspace. Still, Double Attack has its merits in situations for which it was evolved. Say that you're engaging a couple of low bandits with your four-ship flight: if you have another flight with you, it will certainly be overkill to bring them all down. Better then to have the second flight stand off and loiter at altitude while your first flight sorts out the enemy below. Same thing if all you have is a wingman – you might feel confident in managing a lone bandit on your own and judge that your wingman is better employed as lookout against possible wildcards while you deal with your quarry. Should things turn foul, you can always call him down to have a go while you disengage.

LOOSE DEUCE

The pinnacle of tactical evolution is the Loose Deuce doctrine, which nowadays is the only tactic worth discussing at length and to which we will devote all our attention. The fundamental notion here is that every part of the formation is to manoeuvre in such a way as is most profitable at any one time, be it as individual in a team of two, or as a four-ship flight in a squadron of sixteen. The key to Loose Deuce is separation: how to use space and the enemy's predictability to overwhelm him in the least amount of time. Loose Deuce tactics properly employed is a thing of beauty to behold and is indeed an aerial ballet, a game of four-dimensional chess which leaves the unsuspecting or less knowledgeable enemy completely stymied and very soon extremely dead. In the Loose Deuce doctrine, when combat occurs, the normal leader-wingman relationship is dissolved. Instead, you have two equal opportunity fighters who support each other and complement each other, either of which can be either "engaged" or "free". These roles may switch at the blink of an eye depending on the angles involved, and one thing that sets the accomplished pair aside from the less experienced is the instantaneous knowledge and radioed acknowledgment between fighters regarding who is engaged and free at any one time. Henceforward, the tactics discussed all deal with this "Loose Deuce" arrangement. The story below is from a textbook sortie where a wingpair in Dewoitine D.520s engage with team tactics against multiple enemy aircraft:

Formed with Starry after he shot down a 109E low off Montfaucon after an attempted vulch, swept the area clear. Three 109s had been reported in the area earlier. Formed in line abreast climbing east. Spotted con long 6 in pursuit which turned out to be an Emil. I chandelled on him then dragged him east for Starry who got an easy kill. Reformed and headed north in line abreast. Starry spotted con south and somewhat low, immediately went in with Starry in trail. Con turned out to be 109 coming in head-on. I chandelled around him, barrelled for a solution but had to abandon that fight when one other Emil appeared, near co-alt but low on smash. Both were looking at me, well out of range, so again I dragged east. Saw Starry come sweeping in off their high seven o'clock going for the low and far trailer first. He got a pilot kill on that fellow, the other was seen fleeing north on the deck. Reformed to east, then north. We arrived Bertrix and set up shop at 3km, Starry remaining at 5km.

Three cons on the runway with two rolling, so I called Starry down to low orbit. I went in on Stuka heading north but failed my pass grossly by pulling too much lead. The now well alerted bandit hugged the Libin ack for a while, then set out northwest again. Boomed him in three or four passes before setting him afire. Back to the field. Starry engaged an 87 who managed a burning ditch well outside the drome limits. I next went in on 109 and 87 heading out north but missed the 109 who broke at the last moment. I chandelled up and lost vis on the bandit. Meanwhile Starry also went in to north. Shortly afterwards I spotted a con to north at about 2km and engaged, finding Starry in pursuit of two cons near Libin. Starry went for the high 09, I went for the Stuka. Stuka flamed in one pass. Starry took care of the 109 while I engaged a 110 who joined the fight. Pursued him towards Wellin before finally downing him. Starry nailed the 109. Reformed in vicinity of Wellin, spotted con to north assumed to be Starry. Proved to be 110, weaving desperately. Starry beat me to the kill. RTB *to Maubeuge low on ammo. EXCELLENT sortie in all respects, particularly in comms and engagement discipline. Final tally: bmbm 4 kills, starry 5 kills and a 6th attrited (Stuka). Mission duration 1:10.*

Before we go into details however, it should be borne in mind that the Fighting Element, or the Flight, or the Squadron, deployed in combat spread and fighting according to the Loose Deuce doctrine, is ALWAYS but always offensive. There is no such thing as defensive manoeuvring. While the formation may perform guns defence or manoeuvre to avoid an enemy attack, its outlook is ever offensive. Every seemingly defensive move is a transition to attack. Even when disengaging from a fight, the formation is actively attacking or getting ready to attack its pursuers. It should also be borne in mind, again, that individual manoeuvres such as discussed in the ACM section have precious little application in the formation fight. Realise this when looking at the diagrams. In order to make the Loose Deuce work properly the fighters must retain a largely similar energy state: once one of them bogs down and blows his energy by manoeuvring too sharply or engaging in single combat against an enemy, the relationship becomes disjointed and suffers accordingly. In Loose Deuce combat you have to fight in a totally different manner, in a teamwork dimension which is entirely separate from that of the single fighter. Remember that always: keep your energy up, watch how your moves affect that of your partner and that of the enemy, and you cannot go wrong.

18.3 THE BRACKET ATTACK

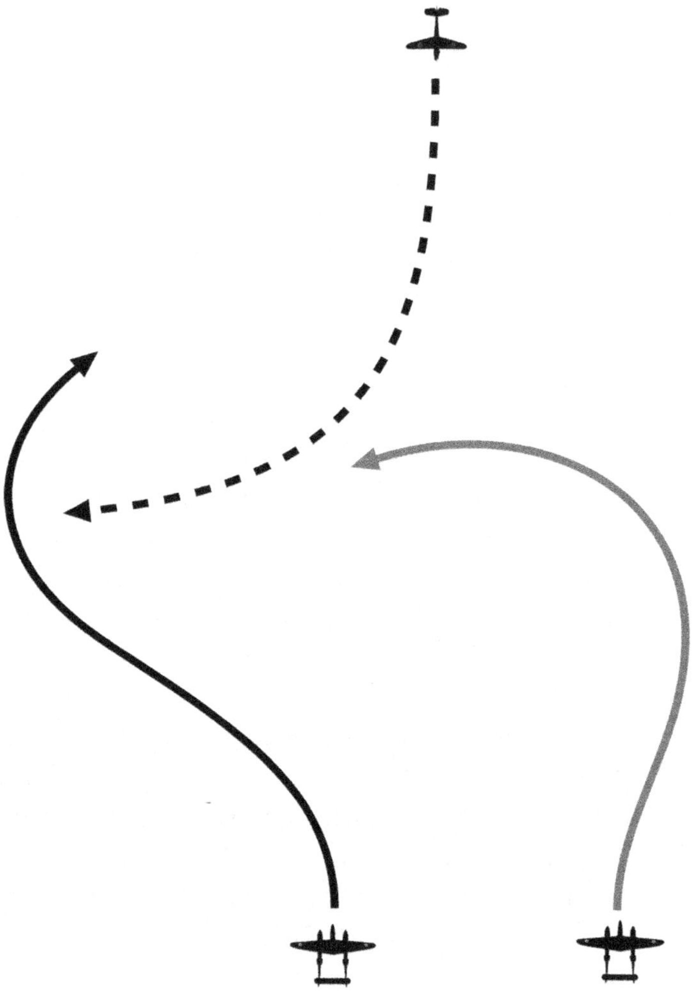

Fig 58. Bracket attack: 2 versus 1

The bracket attack is a bread-and-butter tactic used by the element or larger formation when encountering the enemy in their forward quarter. It epitomizes the philosophy of the "free – engaged" doctrine and clearly shows why and how separation is such a powerful tool. By forcing the enemy to split his attention and engage just one part of the element at any one time (the "engaged" fighter) the "free" fighter gets a clean shot. Throughout the engagement the roles of the free and the engaged fighter switches back and forth depending on which of them has the enemy attention, until the enemy is defeated. The opening move sets the scene: from the cruise spread, the fighters angle out to put the enemy in the middle and then turn into him at the point where energy, closure and range dictates. At that point, if not before, the enemy must choose which of the pair to defend against – he cannot defend against both.

Figure 58 shows (not to scale) the mechanics of the bracket. Shortly after the wingpair opens up separation (on the command "Bracket! Go!") it becomes apparent that the bandit (dashed line) chooses to manoeuvre against the left fighter (black line) – probably because the black fighter was first to yield angles. The bandit reacts out of reflex, and the rest is a foregone conclusion. The black fighter merges as per usual with the bandit, performing guns defence as needed and avoiding the temptation to line up a shot. Meanwhile, the grey fighter (the "free" fighter) has already started his pursuit curve and is closing in on the bandit's six. At this point the bandit is already dead – it's now just a matter of time and gunnery skills on part of the grey fighter. As you can see in the picture, what begins as a pretty even fight is ruthlessly transformed into a turkey shoot.

The notable thing here is that the black fighter makes a point of getting outside of the bandit so as to force the bandit to keep his lift vector on him after the merge – which sets the enemy up in a most conspicuous manner for the free fighter's shot. This is most vital, as the instinct to keep your lift vector on the closest and most immediate threat is overpowering and thus something you can count on happening. Should he have passed the bandit on the inside, the bandit would be turning towards the free fighter, something which would be less desirable as the grey fighter would then face a head-on situation instead of a tailchase.

How do you defend against the bracket? It's pretty easy actually. There are two options, use of which depends on the ranges involved: to blow through, or to line up the wingpair. In the first option shown below (fig 59), the defender fakes an interest in the black fighter but only so long as to see him flush out to the side and to take stock of the real threat: the grey fighter. The defender leaves the black fighter in the dust, forcing him to turn a full 180 degrees, and performs guns defence against the grey fighter's onslaught. The defender should take care to keep the grey fighter in front of his wingline prior to the break so as to require him to complete a full 180 degree turn relative to the start of the combat. The defender, never turning more than 30 degrees off his initial heading, keeps his steam while the wingpair will end up wallowing in his long six.

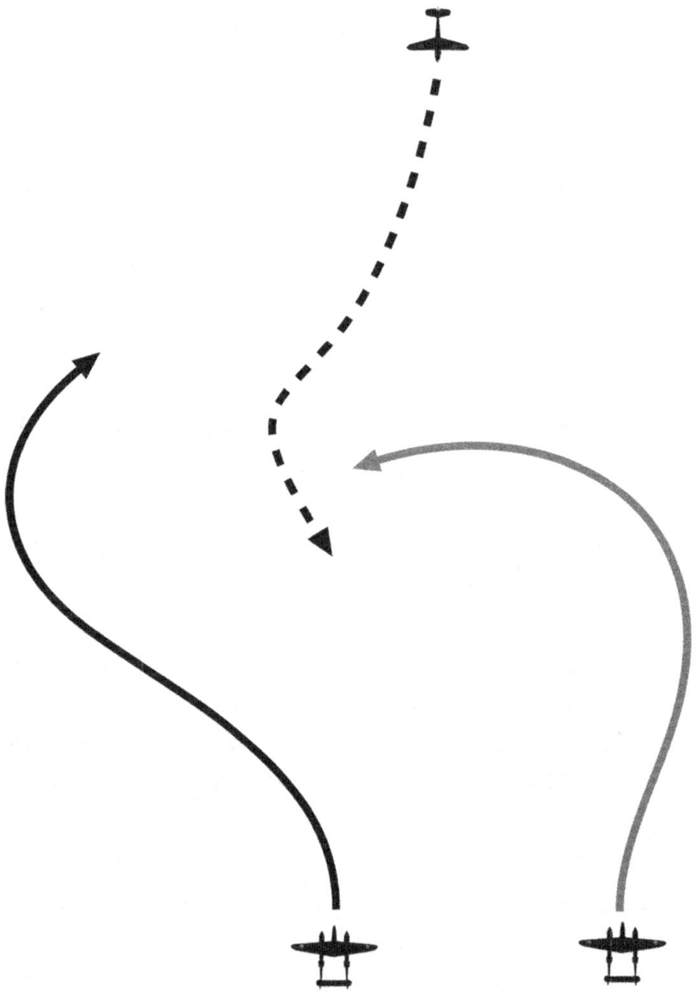

Fig 59. Defending against the bracket: coming through!

The second option, lining up the wingpair as one entity, requires more time and space. Thus, the defender must decide early on, immediately on sighting the pair at extreme range, how to handle the situation. In fig 60, the defender angles out instead of going straight in at the enemy, seeking the "outside" position. In so doing, he forces the wingpair to commit to lead pursuit which in turn forces the grey fighter into a trail, or at the very least, echelon, position to the black fighter. The defender is not allowing the bracket pincers to open and instead reserving the right to decide for himself. In fig 60 the defender wins the outside position and makes a sustained turn into the wingpair, now conveniently grouped up in front of him and deprived of the separation essential for the bracket attack. The defender barrel rolls right past them and continues on, forcing the wingpair into an expensive 180 degree turn that will leave them high and dry.

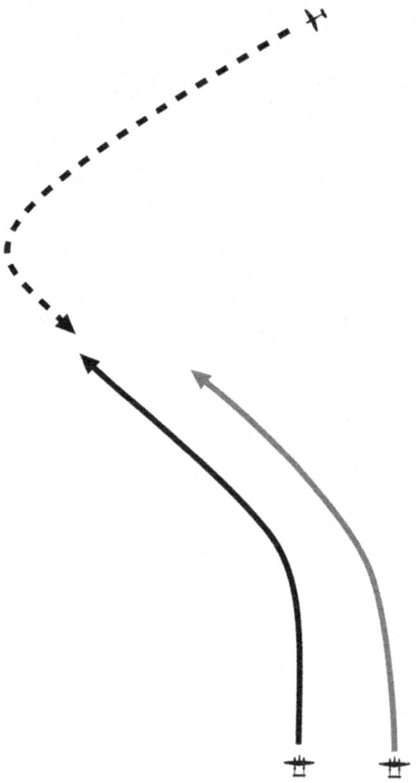

Fig 60. Defending against the bracket: line them up and blow on past.

However, let's say that the wingpair has been around this situation before and have learned how to handle it. In that case, the grey fighter controls the urge to follow his leader and opens up separation instead, restoring the bracket and waiting for the bandit to commit. In figure 61, the black fighter is left in the dust by the defender's near head-on pass, but the grey fighter is coming in hard to close the deal.

As can be seen, the outcome is not such as foregone conclusion as one might think. It all depends on the decisions you make, and whether the opponent is a fool or an artiste. The defender in the scenario presented in figure 61 could have chosen another route: upon seeing the grey fighter loading up on separation he could have performed a closer guns defence with less angles expenditure (not the 90 degree turn as shown but a 30 degree barrel roll instead) and kept on trucking left out of the picture. In that case the grey fighter would have yielded angles and distance for no gain, having been flushed right out of the fight as it were.

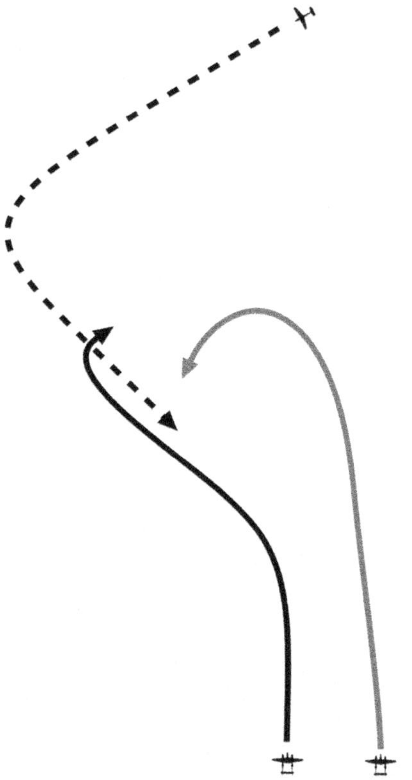

Fig 61. The bracket restored: the defender is in a world of hurt.

Now, the various bracket manoeuvres are just as applicable for a flight of two elements or a squadron of four flights engaging a single enemy or an enemy formation which flies as a single group. Once the enemy formation also splits up however, the bracket becomes more interesting.

In figure 62 the standard bracket shown previously is not materially different from the 2 vs 1 situation than when applied against another pair. The bandit pair (operating on the welded wing doctrine) goes as one toward the black fighter who defeats their passes in sequence, allowing the free fighter to bore in at his leisure. While this situation is not uncommon, it is however not completely certain that the grey fighter will have time and space sufficient to blast them both before the fight degenerates to a melee. It is far more common for the bandit pair to split up as well, as in figure 63.

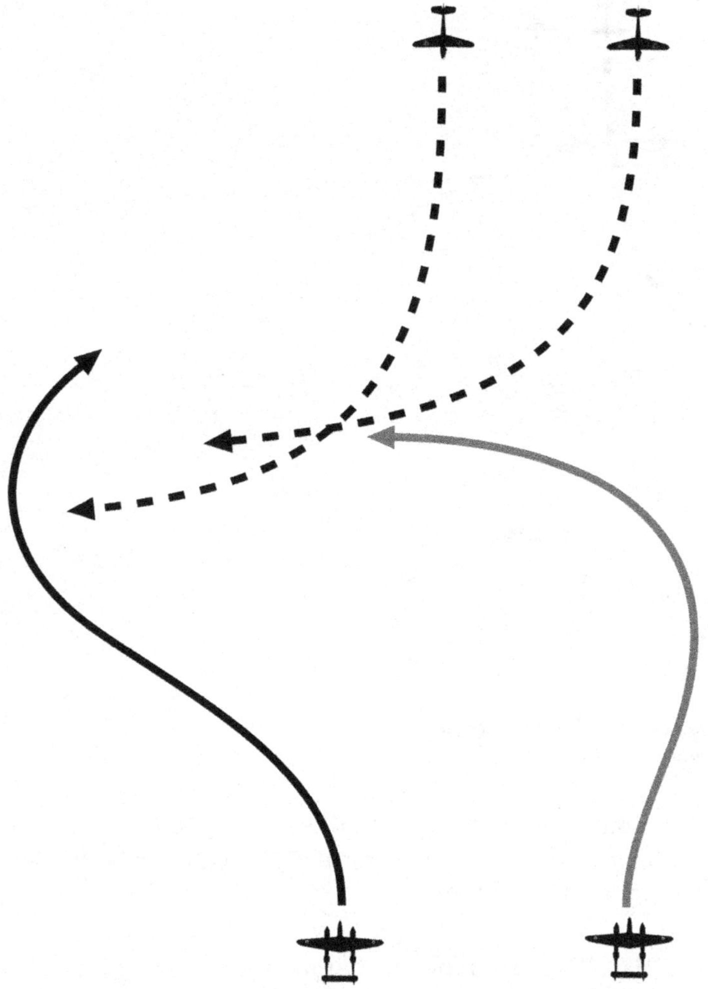

Fig 62. Bracket attack: 2 versus 2, bandits committing as one.

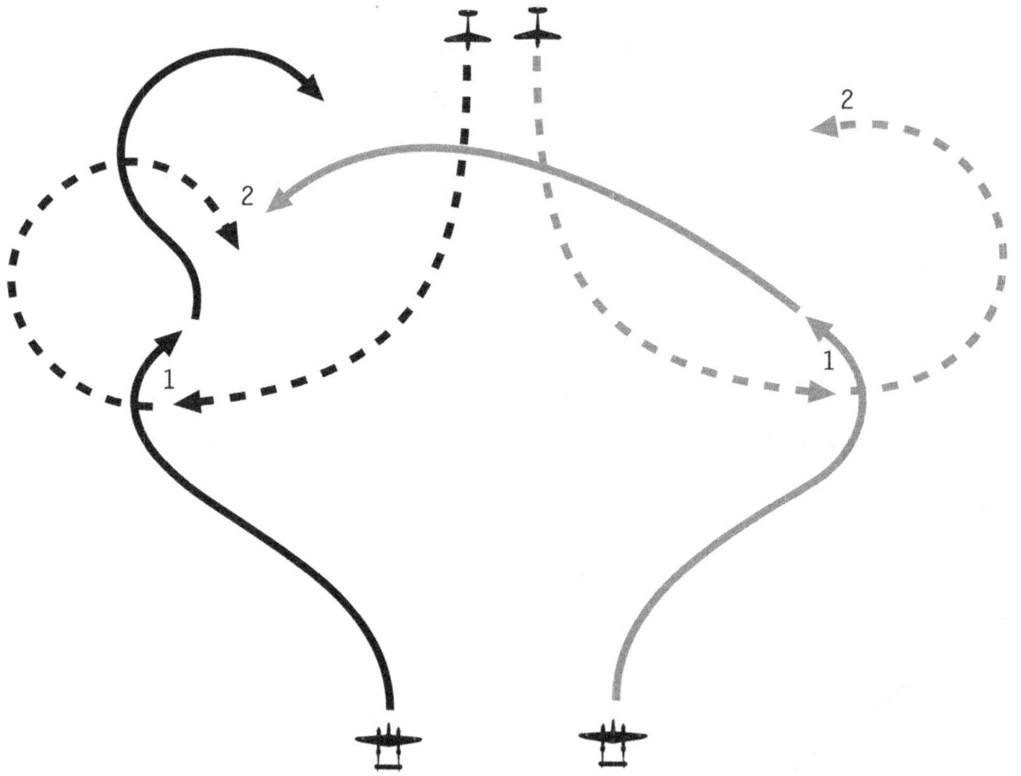

Fig 63. Bracket attack: 2 versus 2, bandits split up. At time 2 the left bandit has been forced, twice, to commit against the black fighter, setting the stage for the grey fighter who has left the second bandit well behind and unable to help his mate.

Here, the same mechanics are at play. The difference is that instead of going into two separate single combats, the home team (solid lines) concentrates on knocking out the enemy in sequence by ganging up on one while leaving the other high and dry. The technique is similar to that of the 2v1 engagement: both fighters merge on the outside of the enemy so that the home team's guns defence turns are made toward each other. By default the enemy will strive to keep their lift vectors on their respective targets, at a steep energy cost, thereby forcing them outward and creating a separation which is too great to close before the home team has annihilated the chosen target. The key here is for the home team to select and communicate between each other which of the enemy to single out for immediate destruction. Ideally, the selected target is defeated in the opening stage of the combat, and with one of the badguys out of the way, it's a simple matter to wrap things up and deal with his wingman as in the 2 vs. 1 bracket.

Again, this tactic applies also to greater formations (flight, squadron, group) in the same manner versus enemy formations of virtually any size. However, if the force disparity is too great, the element or flight had better blow right through and fight another day – unless of course a significant energy advantage is at hand.

The same theory of separation can be applied in the vertical plane with the same cheerfully predictable results: the High Bracket. Here, the lead fighter calls out the manoeuvre and starts a sustained zoom, slightly offset to keep the bandit in view, forcing the badguy (dashed line) to choose between taking on the high or the low threat. Most pilots act instinctively and go for the high threat, shown below, but either is fine. What happens is that the free fighter gets a million-dollar shot against the bandit's low cold six, at a particularly destructive angle off too. Should the free fighter miss however, which is rather unlikely, it's not overly catastrophic as it's a rather simple matter for the engaged fighter to spiral around the bandit's lift vector and dodge his shot. In the case the bandit goes for the co-altitude wingman, he performs a simple guns defence while the high fighter uses his gravity assist to fall in on the bandit. Piece of cake!

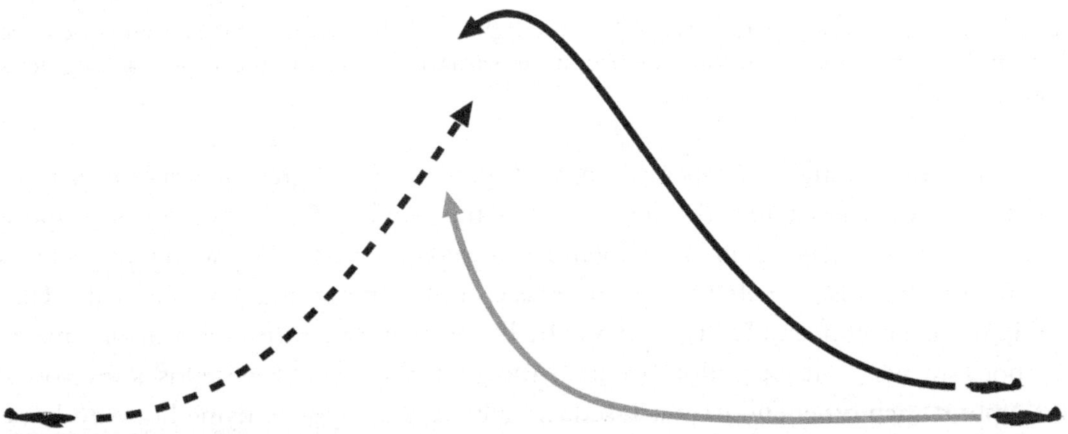

Fig 64. High Bracket (not to scale).

Similarly, separation can be created by sending one fighter below for a Low Bracket – the engaged fighter makes a standard guns defence, outside break, sending the bandit careening outward and a prime target for the now zooming low fighter. It is quite unlikely for the bandit to go for the low fighter initially – it is rather more likely that the bandit will blow through and not accept the fight at all, in which case it's a slim matter for the free fighter to use the overhead energy and roll up on his extending six o'clock. All of this will become apparent well before the merge, so there's plenty of time (2-3 seconds!) to detect and prepare for the upcoming move. As in the other variants described above, the discerning defender can rather easily deflect the high and low brackets simply by not committing fully to either wingman. As long as the defender times his moves appropriately and takes care to defeat each threat before they grow past 45 degrees of deflection, he can fake and parry each move in turn. Key to defending against the bracket is to keep going at a good clip, to pass each threat well inside the bandit turn radius yet outside of a head-on shot opportunity, to avoid pulling more than 30 degrees turns and to use barrel rolls rather than stodgy old nose-over-horizon break turns.

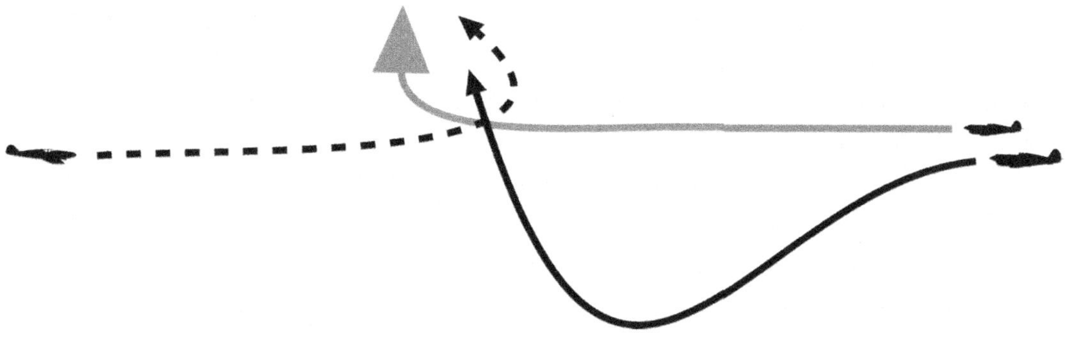

Fig 65. Low Bracket (not to scale). The grey fighter, farthest away, makes a guns defence turn towards the camera, flushing the bandit outwards and unable to meet the black fighter's low six attack.

The only thing that can really mess things up is the great temptation a team pilot feels to manoeuvre individually against the bandit. When engaging in a loose deuce fight the main thing to remember is separation. Once the fighters closes separation they lose the ability to support each other and to attack sequentially. That said, the separation should ideally never be bigger than one turn radius or one major manoeuvre away – if separation becomes too great the team likewise loses the ability to support each other and to attack sequentially. It takes a lot of flying hours to hone this team, a lot of communication and a lot of post-action discussion to fine tune moves and responses. Don't expect any odd team to succeed with loose deuce tactics from day one.

18.4 TRAILING ATTACK

When the element has the drop on an unsuspecting low bandit it is auspicious to open up separation "artificially" by having the wingman "hang back" a few seconds before following the leader. If the leader misses, the wingman gets a clean planform shot at the breaking enemy or an opportunity to nail the startled bandit as he recovers his wits and concentrates on the disengaging leader. It's quite uncomplicated really – the only thing to keep in mind is that the wingman should duplicate his leaders' attack, i.e. if the leader goes for a deep six pursuit, so should the wingman, in order to not cut him off or allow the bandit to gain visual contact. After the attack, should the bandit still not be defeated, the home team should make their next moves in opposite directions, i.e. once the wingman concludes his attack, he zooms away from the leader to create the necessary separation for another sequential attack on the bandit.

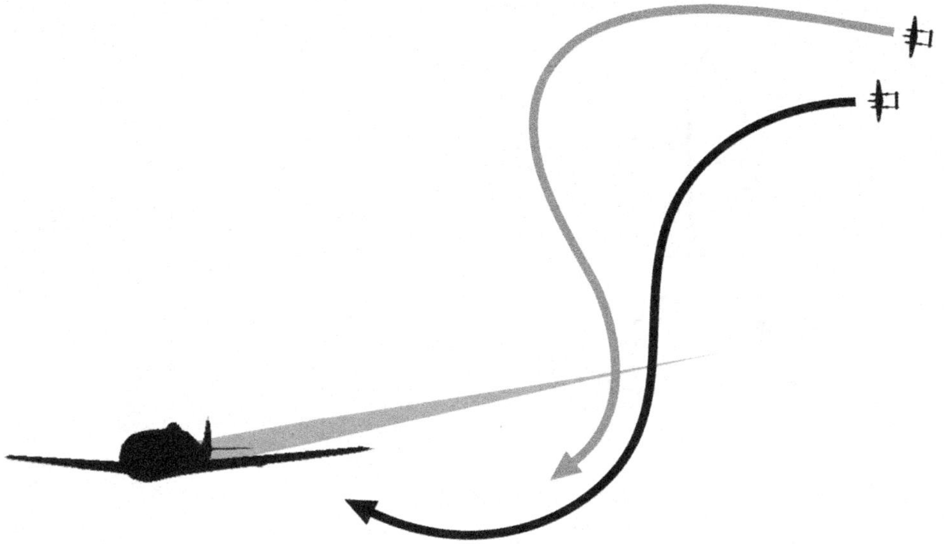

Fig 66. Trailing attack. The wingman (grey) delays his attack run a few seconds to create separation.

18.5 CROSS SPLIT

The Cross Split is a great example of how the fighting wingpair is always on the offensive, even when it may seem like they're on the defensive. While the team starts out with a bandit on their six, the transition to the attack is swift and merciless. Key to the manoeuvre is to detect and act against the threat before the bandit draws into guns range, which, as long as you're in line abreast, is a simple matter. The team makes a sustained turn toward each other, observing the enemy. The turn creates separation, forcing the bandit to make a fast choice of whom to attack – he cannot attack both at the same time. This, the bandit's choice, will become apparent in the turn, or, if the engagement starts at long range, as range decreases. If the manoeuvre is started at extreme range, you treat it as you would a standard bracket attack – make sure to maintain a sufficient separation between fighters so as to force the bandit to make a choice rather than to attack both fighters as one. Once the bandit's intent becomes clear, the engaged fighter performs guns defence as in the bracket (taking care to pass the bandit on the outside so as to send him careening outward) while the free fighter motors up on his six o'clock. From there on it's a short distance to the endgame. If endgame does not occur and the situation allows for engaged manoeuvring, make sure to continue working the bandit from different directions in order to make him break or overtax his SA. In other words, maintain separation.

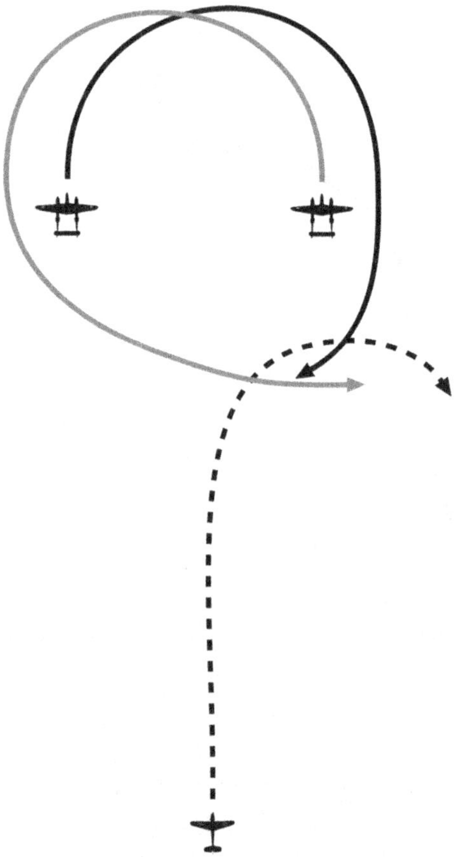

Fig 67. Cross Split. Separation is employed at the outset, throughout the turn to confirm who the bandit goes for, and following the free fighter's shot.

18.6 SANDWICH

Separation is useful however you employ it. If you had rather not make a full reversal (such as in the Cross Split) into the enemy for reasons of avoiding a particularly bandit-infested area, forcing the enemy to commit to one of the fighters always gives the second fighter a free shot. Determining who the enemy is gunning for is imperative: as soon as the bandit is seen to commit to the right-hand fighter, the free fighter (grey) swings in with lead pursuit.

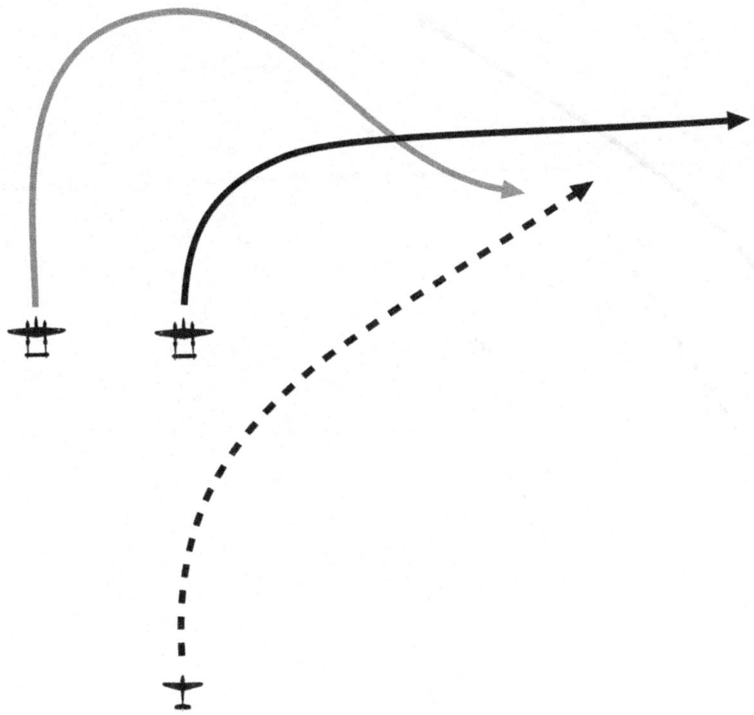

Fig 68. Sandwich, or Defensive Split, against a high-closure bandit.

The free fighter must time his reverse carefully, preferably so as the bandit shows him his belly at the time of the reversal. If he reverses too soon the bandit may switch targets, if he reverses too late the engaged fighter is put at risk. Generally speaking this manoeuvre should be initiated well out of guns range and with moderate bandit closure so as to give the wingpair sufficient reaction time. These illustrations are not to scale with regard to range – if at any point the engaged fighter is inside enemy guns range, he must perform guns defence.

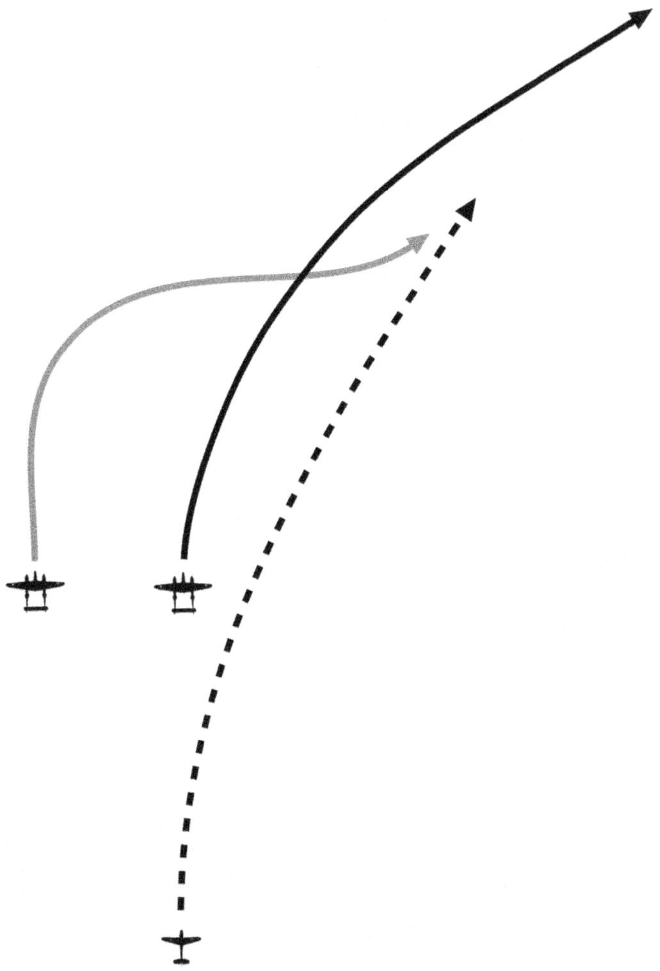

Fig 69. Same manoeuvre against a bandit with low closure: the setup can be made more leisurely at a lesser cost in energy.

The Sandwich is dangerous inasmuch as it relies on baiting the enemy with what appears to be an easy kill – the engaged fighter is banking on the fact that his wingman is on top of the situation and can close the distance before the enemy does. Therefore, always use the Sandwich with caution, and when closure and the ranges involved are manageable.

18.7 HALF SPLIT

The Half Split is virtually identical to the Sandwich – but here the bandit goes for the extending wingman instead. Timing the reversal is the critical part here – if the splitting team member reverses too soon, he will have a hard time lining up the bandit and may even slide in in front of him, clearly not the desirable thing. Again, this manoeuvre is best employed when ranges and closure are manageable – if the bandit is too close to guns range, engaged manoeuvring is preferable (e.g. the cross-split).

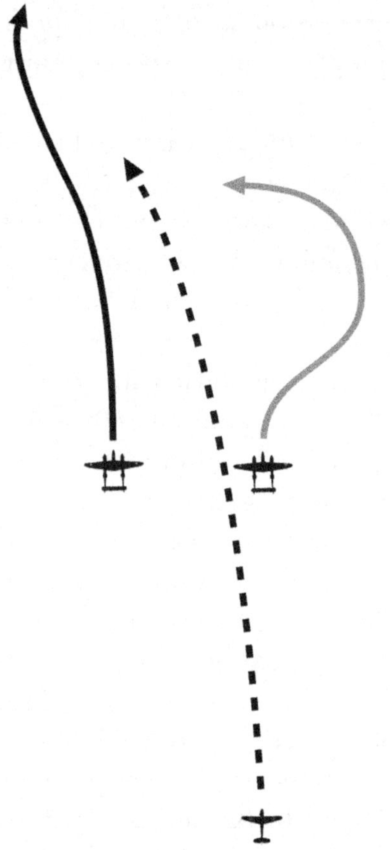

Fig 70. Half Split, with the bandit going for the extending defender.

18.8 THACH WEAVE

Named after its inventor USN Commander Jimmy Thach, this manoeuvre is best described as a team scissors – reciprocating S-turns - with the enemy at or very close to guns range. It can only be executed by formations of at least element size, you cannot perform it on your own, and you need to have practiced it a lot too.

The Thach Weave, or Beam Defence as it was initially known, is an excellent and extremely unnerving (for the enemy!) defence against single or multiple enemy aircraft in a situation where your own craft are inferior or where your main concern is to disengage promptly. Note that you absolutely must be flying line abreast to perform the Thach Weave! An important point is to make your initial turn 45 degrees (or thereabouts) off the general heading. If you're forced to turn more than 90 degrees (i.e. 180 degrees for the subsequent turns) relative to your general heading, you're becoming stationary and about to be engaged in a dangerous turnfight – and then the weave stops working. Maintaining forward progress is therefore incredibly important. Every time the scissors "closes" the free fighter gets a very brief high angle off snapshot – with plenty of deflection and a judicious outlay of ammunition, preferably in an unloaded state, the bandit, whose entire planform is exposed, must fly through a curtain of fire.

The Weave can be used by formations as well (i.e. with elements or flights where individual planes are shown in the illustration), and both at high and low airspeed. At higher speeds and longer ranges, the weave must open up and become flatter, i.e. with less marked turns and with greater separation. In this manner, the team, both fighters hotly pursued, can "shoot bandits off each other's tail" almost indefinitely.

What if you're not in line abreast but in trail (wingman following the leader) when you're hotfooting it out of enemy territory? Simple. Picture yourself 1000 yards behind your leader. Not far behind is the evil enemy, though still out of guns range. You're pedalling as hard as you can, running scared, knowing that he'll shortly overhaul you and gun you down! And then you think of the Thach Weave, why not try that one for a change? Surely it must be better than being shot down screaming like a girl? So, you tell the front guy to angle out some 30 degrees while you too angle out a similar amount, to create lateral separation between the two of you. Once well spread out, get back on the previous common heading. Now the enemy is forced to commit to one of you, leaving the other free. Hey, all of a sudden you've reduced the threat to half your force and created an opportunity to counterattack! You are now parallel with your wingman but several thousand yards apart, the bandit back there is straining to get in range. Now tell your friend to make a sustained (gentle) turn towards you – you do the same, turn toward your wingman. The bandit attempts to cut the circle and close the distance but there aren't enough angles! And here comes your wingman, caroming in with a nice 30 degree lead on the bandit. Frrrrrrrrrrrrt!

It should also be easy to see by now that the less distance you have between the two arms of the scissor the easier it is for the free fighter to get into a rolling fight with the enemy. In other words, it's fairly uncomplicated for the free fighter to pitch up a bit, roll inverted and then haul down in the bandit's stern quarter for a low-deflection shot.

The real utility with the Weave is in the few-against-many fight, provided you're in aircraft that can take a few hits and still keep ticking. If you're four against fourteen you cannot afford to saddle up even for an instant. Picture the Weave then where both the arms of the scissor are under attack. In order to survive you must jink, bob and weave somewhat until you get to the point where the scissors close, where the other half of your team can close the distance and gun down the enemy.

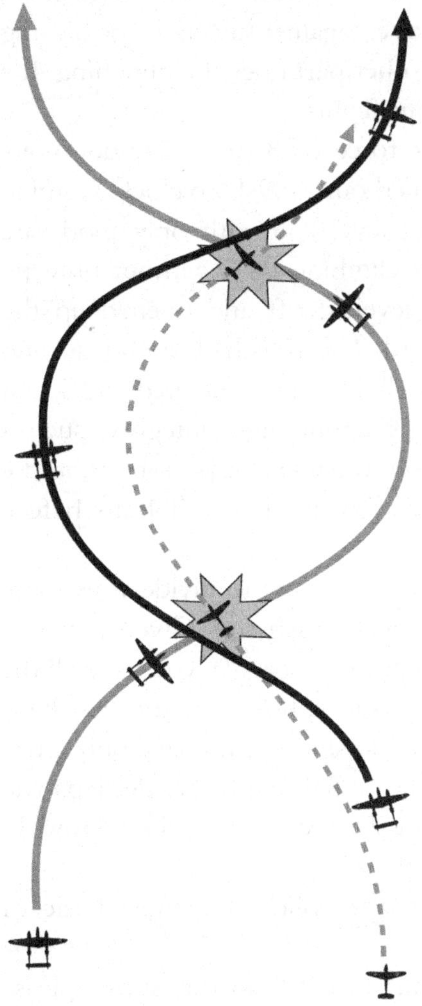

Fig 71. Thach Weave. The evil bandit (dashed line) begins the engagement out of guns range and attempts to cut off the engaged fighter (black line). The free fighter puts paid to that endeavour!

A WORD ON BIG FORMATIONS

The manoeuvers described above are best employed by formations numbering no more than 2-4 aircraft. Once you have a greater formation it becomes increasingly difficult to execute with any degree of success, say, a Thach Weave or a Half Split. While an element (2 fighters) or a division (4 fighters) may happen into a situation that warrants such manoeuvers, a bigger formation will have sufficient numbers to enforce the decision without them. For instance, you wouldn't engage your 16-ship formation in a massive drag-and-bag against a piddly pair of bandits: you would instead commit only so many fighters as are required to defeat the numerically inferior enemy while keeping the remainder in reserve at altitude. On the other hand, a major formation may use the same kind of moves against another equally big formation, i.e. use part of its force as bait while the other part does the punching. It's all in the use of separation, and in the integrity of your flights.

I remember very well an occasion when our force of 8 or so Thunderbolts were climbing east over Flushing when someone called out "109s! 6 o'clock! Coming in!" and verily, there came two 109s moseying in, level and with only moderate closure. What to do? You can't easily reverse eight climbing P47s without blowing the whole formation to smithereens. Instead, we levelled off and opened up the formation, splitting four to port and four to starboard. The 109s had to choose, and as they chose, the free flight rolled back into them and generated sufficient intimidation to cause them to abandon the chase. Without flight integrity, such a manoeuvre would not have been possible, and without the line abreast formation the 109s would have closed unseen and knocked down a fair number of T-bolts before the rest caught on.

If you do happen to head out in massive force, make sure to divide your force in divisions of two, four and eight aircraft as previously discussed. Given that you share voice comms and fly as a disciplined force rather than a gaggle, you have all the opportunity in the world to dominate the sky. More usually, however, you may have been elected leader over a bunch of fighters in a scenario setting, and have only a few minutes at your disposal to array this motley crew into a fighting force. Perhaps you have advertised a mission and a large crowd has turned up, eager to follow. How do manage that?

Make things as brief and simple as you can, and follow the most basic of routines:

Establish you leadership in no uncertain terms: "I will lead this show, please listen up. I will answer questions later."

Designate flight leaders: "We will fly three flights, each consisting four fighters. I will lead A flight. Boggins will lead B flight. Charlie will lead C flight." It is of course best to assign known entities as flight leaders, such as you know are able to hold formation and able to act on their initiative should circumstances warrant it. Ideally you should be on voice comms with the flight leaders.

Sort pilots into flights: "Acne, Abrams and Arse, you're with me. Biff, Bumpy and Blowhard, you're with Boggins. Cutie, Clinch and Crispy, you're with Charlie. All, please taxi up to your leaders now." If there are latecomers, distribute them between the existing flights, or add subsequent flights as required. A flight can be of any size, so even if you have an unwieldy 40-strong force to contend with, you may still divide them equally into 3 or 4 easily managed groups. All the pilots need to do is pay attention, and stick to their designated leader.

Describe the mission: "Our mission is Fighter Sweep. We will take off on my command and make one full orbit over the field at 2000 feet. On my command, we will head out EAST, climbing at 75% throttle. Use whatever throttle necessary to format with your leader. Flights will be in line abreast: A flight in the middle, B flight to left, C flight to right. Keep about 500 meters between flights. Pilots, stick close to your leader at all times. We will level off at 20,000 feet and sweep to Targetburg. Once there, we will turn NORTH to Middelstadt and then WEST to Halfville before returning home. When we engage, the spotting flight will go first into the attack unless another flight is better placed. If you're separated along the way, make best time to these waypoints and orbit there at 20K. If unable to rejoin, make best time home. Questions?"

This sample briefing, requiring no more than a few minutes to conclude, contains every scrap of intel the flight members need. Even so, only half of what you've just said will filter through most brains, and fewer still of your pilots will have patience and wits enough to stick with the plan – trust me, I know what it's like. Sometimes you will only have a single dependable fellow as your second flight leader, and an overhead of eager pilots. In such case, form two flights and move them out in double column (two parallel line astern formations). If you're down on luck and have no second flight leader at all, designate one on random and go with the double column anyway. However, instead of having your second flight format on your own, you should format on it. In such a manner you have at least ensured a small measure of tactical flexibility and visual overwatch. Realise that the less you know about the individuals comprising your force, the less demands you can make.

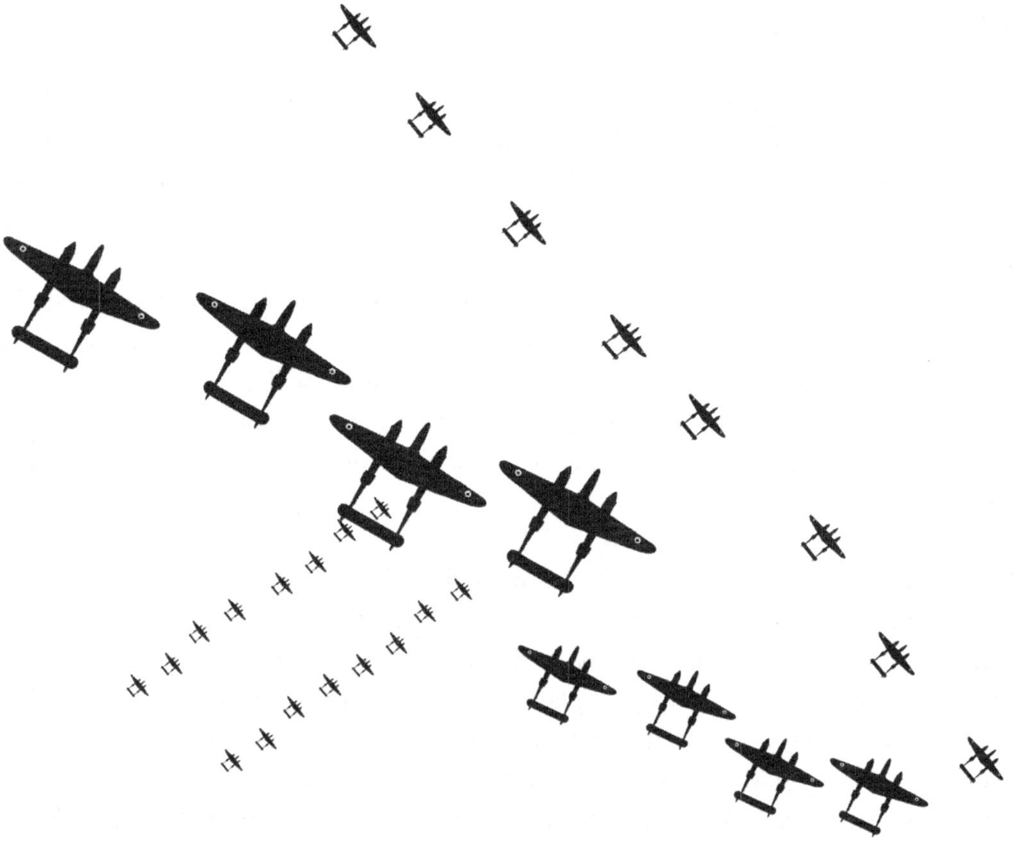

Fig 72. Two high flights and a squadron in wide line abreast covering a double column.

18.9 ENGAGEMENT AND DISENGAGEMENT

"Locating Trimcz near Herbeaumont, I help him disengage from two 110s and commence the homeward leg in the vicinity of Mouzon. The group returns singly or in pairs from the area. Woolef makes an early RTB after sustaining wounds from a rear gunner, Hammered suffers a collision after about an hour on station, Duff is lost in action, but the remainder returns more or less safely to home plate. Group scores 25 confirmed kills for 2 losses in a mission lasting 2 hrs 15 minutes. Pretty good return ratio."

WWIIOL, February 2003

As stated before, it's easy to get INTO a fight but all the more difficult to get OUT of it. To that end, it behoves the astute flight commander to keep the exit in mind even before the fight is joined. Adhering strictly to the "one pass, haul ass" principle helps immeasurably, as does having a team which recognises this principle. It only takes one flight member to muck it all up – everyone must be on the ball and stick to the rules. Time after time you will come up on formations that know very well how to get into the fight but knows nothing about getting out of it – in very short order the formation will become bogged down and break down into a series of individual combats where no one is supporting the other.

THE FORMATION ENGAGES

Having been in innumerable formation versus formation engagements, the thing that strikes me as the most significant is how extremely small the window of opportunity usually is and how little it is that separates a good engagement from a bad one: the formation leader must make a split-second decision, communicate his orders, preferably communicate with the force commander, and manoeuvre himself and his formation – all within the space of a few seconds.

"I was in Boomer's flight, flying second-in-tail [we didn't know any better] of the second column, having just reached 33,000 ft somewhere in the Hannover area – our Mustangs were wallowing at that altitude, although cruising at 400 knots our crates were balancing on a stall – indicated air speed around 180 knots. One false move, one moment's inattention and you'd be stranded hopelessly behind. Everyone was hollering at the laggards: "Close up! Close up!" Just then, off our long 10 o'clock, appeared a ragged bunch of contacts, growing rapidly bigger. "Cons! 9 o'clock!" – "Bandits?" – "Red Flight?" – "Focke-Wulfs!" – "Bandits 9'clock!" We had barely time to become afraid before they swept past us, no more than 500 foot above and as much in front, streaking from our left to our right, never wavering a foot from their heading. Supposedly they were as surprised as we were, for they turned as little as we did, hardly even waggled their wings. And even if we had turned we wouldn't have had the slightest chance of catching up with them."

Operation Pointblank, WarBirds 1997

Things happen incredibly fast in the air. One second you may be in the grandest of formations and in the next you may be scattered to the winds, everyone frantically trying to get a grip. How well you cope with opportunity and surprise is the true mark of your proficiency both as a group and as individuals. Know that a chain is never stronger than its weakest link: one guy lagging behind may put the whole group in jeopardy, one guy blowing a pass or getting bogged down low and slow may kill the whole formation through his ineptitude. While it's bad form to leave a formation member in the lurch, the leader must consider what's best for the group as a whole: sometimes this means sacrificing the weak to save the strong. It's indeed a frightening thing to be left alone in a patently hostile sky, particularly after having enjoyed the relative security of a formation. At once you feel naked, vulnerable and slow – a target for anyone who happens to look crosswise at you. As a formation leader you need to have contingency plans for such occurrences: a "bugout vector", a Rally Point, or simply an order to return to base for the offender. At times you may have the luxury to bail out the laggard or the critically engaged, because you don't have a particular mission or can afford to shed altitude. At other times you may have a pressing engagement with a bomber box or are getting close to bingo fuel[1] and must choose whether to save one guy or to fulfil your primary mission. These decisions are never easy, but they can be as long as your formation members are made aware of your Rules Of Engagement.

"We were heading southwest, at 28-30 K, deployed in a ragged bunch as most others though supposedly in two sections, looking for a gaggle of Spits reported to be in the Ramsgate area. Horrido! Tally twelve plus bandits, low to front and heading northish – here we go! Our flight leader winged over into the attack, the rest following suit. I made a sweeping attack towards the middle of the gaggle, they were in a lazy turn to left, apparently at odds as to what to do. I saw strikes on one, then he broke hard to left, breaking formation. I zoomed up, seeing most of my flight members chasing the bandits lower, lower… lower, going into single combats, shattering formation integrity. I stood the aircraft on its wing, trying to stay above the action, discern the general flow We never reformed as a unit after that."

Blitzkrieg! Scenario, WarBirds 1999

When a formation doesn't have an engagement doctrine other than to "mix it up and see what happens", catastrophes small and big are bound to happen. You, as a formation leader, must instruct your pilots and make sure that they understand the merits of formation combat. Typically, as soon as a single pilot forgets or disregards formation combat practices and engages in single combat, the formation is on the slippery slope of piecemeal destruction. The leader absolutely must enforce discipline, lest he won't have a formation at all. That said, it's quite over the top to expect an ad-hoc formation to perform as well as a battle-honed group, so moderating one's expectations and thereby adjusting to the actual capabilities of the ad-hoc formation rather than pursuing the ideal, is probably the difference between an effective ad-hoc group and an ineffective one. Or, in other words, the less the collective experience, the more basic (as in more disciplined and more security-minded) and brief the engagement must be. Most of the time you will do pretty well as long as you have a single good pilot to act as your second-in-command and leader of your second flight: divide the rest between your two flights and tell them to follow the leaders no matter what.

At other times you will have the distinct pleasure of setting up a near-scientific bounce with plenty of time to consider your approach and team tactics. There's nothing quite as rewarding as having a whole enemy formation to deal with from a position of absolute advantage – savour them, for these engagements will stick to your mind forever. I recall several such engagements and count myself lucky to have

[1] Fuel sufficient to return to base

had the opportunity to experience them. The feeling of having absolute control over a potentially sticky situation, and coming out smelling of roses, is indeed exceptional.

"Climbing out over the Owen Stanleys in our Iron Dogs (P39s), flights in line abreast, we turned north to sweep past Dobodura at 25,000 ft. A flight of Zekes came to greet us: we blew right through them. We could see them literally stop in the air, expecting us to turn around and give fight. No Sir! Another flight was encountered, high on our 11, now sweeping in on our long six and gaining fast. I gave the order: "1st flight, keep going, 2nd flight gentle turn northeast, follow me!". The zeke-gaggle all went after 1st flight. Seeing them cut the distance I told my flight to stick close and broke into them, cutting them off as it were. 1st flight was about to get chopped up! We closed the range, 1st flight now in a state of panic, diving gently to pick up speed, shooting as we came. One, two zekes went down with 37mm hits and then we were through them, catching up with 1st flight. We didn't stop to look back but made poste haste for Port Moresby to refuel and rearm. We didn't lose a single kite in that encounter, but some were damaged and most shook up."

New Guinea Scenario, WarBirds 2000

In the engagement above, a less experienced leader would most certainly have lost control over his group and been the first to wade headlong into the first encounter. The outcome would have been quite predictable: while the formation MIGHT have scored a few kills (the Zeke's being more than a little fragile) one can easily imagine how the formation would have been forced down to the deck and turned to death in penny packets. A few lucky survivors on the fringe of the fight might make good their escape, but the formation would be no more and would therefore have failed abysmally in its mission. I've seen it happen a hundred times. As it were, we opened up separation between flights and watched closely which one the enemy would commit to, then sandwiched them by closing separation. This high-speed half split requires excellent timing and coordination, and we were lucky to have it that day.

THE FORMATION DISENGAGES

"As I gave my 110 a last squirt I heard Hammered say "I think it's time to boogie", and indeed it was. "Yes, let's disengage south". Nose down, roll and roll again, out south. I picked up Hammered: he had two 109s and a 110 in trail. No big deal! Picking up speed, I came in hard and forced them all to break. The 110 and one of the 109s wouldn't let go though. As I streaked past, Hammered zoomed to scrape them off my six. Seeing them break again to his attack, I reversed to clear Hammered's tail. Thus we cleared them all away, downing two, and returned safely to base."

Disengaging from the Metz area, WWIIOL 2003

The key to disengaging is to remain in motion. This sounds awfully trite – after all, the planes are moving aren't they? Nevertheless, a formation can easily become more or less stationary in an area with individual pilots struggling to overcome the enemy in a series of uncoordinated single combats. The formation must always be kept moving as a body, and if it does become stationary the leader must recognise this and order an immediate disengagement in a given heading. When the disengagement order is given it's up to each pilot to untangle himself and proceed without delay in the designated heading, all the while defending himself and helping his team mates to extricate themselves. He who turns back to give battle and he who stays behind in the vain hope of getting just one more kill, will surely delay the disengagement and make it all the more difficult.

Once a few fighters have exfiltrated they should regain energy and positional advantage, ready to help clear off pursuers from extending friends. Once that happens the formation is again moving and employing formation tactics or drag&bag, sandwich etc to regain its integrity. A successful disengagement battle is as rewarding as a frontal engagement and often scores the easiest victories. The difficult thing is to make the breakaway at the right time – in general, it's always better to retreat hard on the heels of a great attack. As the enemy goes defensive or tries to regroup, disengage!

Trying to extricate oneself from an engagement already gone sour is exceedingly difficult not merely because the enemy is already on your six in guns range but because you're so obviously trying to get out of the fight. The enemy will read this as a defeat on your part and will be all the more motivated to gun you down. Thus, start the disengagement when the enemy is reeling. Sometimes you only get to make one proper attack before having to disengage, so make that attack telling.

The disengagement should be made either toward a previously arranged Rally Point, or, if the RP is compromised or simply out of reach, in a cardinal heading perpendicularly away from the most likely enemy avenues. If the enemy expects you to run for home, run the other way. If the enemy is coming from the south, head east or west. If the enemy is usually at medium to high altitude, exit in the weeds.

Depending on the proficiency of your enemy and your own skill, you will most likely lose half of the enemy force at the time of the breakaway. The remainder will chase you, possibly forcing you to take heavy evasive action and to reverse in aid of flight members. The most important thing here is to keep the disengagement going, to get well away from the bulk of enemy, and to get back into a tactically sound formation without delay (i.e. line abreast). Adjust your disengagement vector 45-90

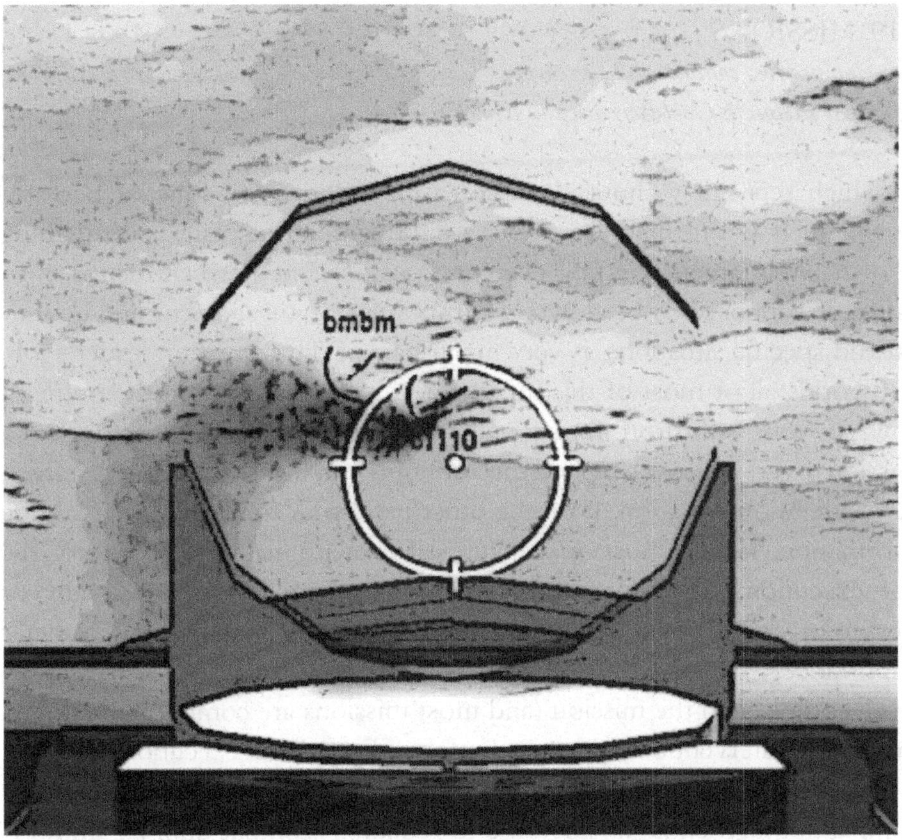

Fig 73. The wingman view of another victory for the author.

degrees away from the initial bugout heading (that which has been communicated to the enemy gaggle by your pursuers!) when you've successfully detached yourself from the main enemy concentration. Eventually, through team tactics, you will reduce the pursuers to a handful, all the while regaining your situational awareness and formation integrity, and at that time it should be obvious even to them that they've been led down the garden path. They started out as pursuers, thinking they had an easy kill coming, now they're nothing but meat on a stick. Reverse and hack them down!

CHAPTER 19. MISSIONS

"No battle plan ever survives contact with the enemy."
Field Marshal Helmuth Carl Bernard von Moltke

A formation which sets out without a mission isn't likely to last long or perform particularly well, much as a formation without a fighting doctrine really isn't anything but a bunch of hapless targets. Getting everyone on the same page and clued in to how the mission is supposed to be flown, what its rules of engagement are, how to act in general and specific situations, is very much part of the formation leader's job. In the best of worlds all or most of this information is disseminated in the preflight briefing. Now, if your squadron functions well already and have performed a particular mission over and over again, this briefing can be short indeed: target, climbout vector, flight organisation. If on the other hand your formation is an ad-hoc beast, the formation leader must make himself exceptionally clear as to the particulars of execution. And since the majority of online pilots are totally or somewhat new to organised (as in disciplined) flight, the briefing takes on a character of utmost importance.

The more complicated the mission (and most missions are complicated enough even without enemy intervention!) the more uncomplicated the execution must be. Just getting people to understand that they must stick together and not get lost in individual combat is tough enough: The urge to remain in combat until all ammo is shot off or all enemy downed is incredibly strong, even though it generally comes at the price of being shot out of the sky before either of the previous occur. Resist this urge in yourself and work like a devil to make formatting and combat as uncomplicated as possible. In other words, fly straight and level without unnecessary rolls and weaves, and limit yourself and the formation to single passes on the enemy. Once you start turning the formation is shot, its integrity blown away.

Regardless of the mission, go through these points in your briefing:

- Airdrome to launch from, and time to launch.
- Platform and loadout.
- Mission type and target.
- Routes and altitudes for ingress and egress.
- Rally Points (locality or landmark).
- Flight organisation in flights and elements.
- Formation, power settings and rules of engagement.

When everyone's sorted out, proceed to launch by forming up on the runway in elements and flights. The few minutes spent taxiing and arranging yourself on the runway are well invested, as it is a lot tougher to form up in the air. When properly arranged, state the initial heading after takeoff and whether you'll make a left or right turnout onto the heading (unless you're already formed on the initial heading). Power up, countdown, and roll as a body. A well-oiled squadron can make it through this process in a couple of minutes whereas an ad-hoc group may require anything from 10-20 minutes simply getting organised. Nevertheless, go through with it! Don't take the easy way out by launching as is – it will come back to bite you in the posterior right away. Another "don't" is to spend precious fuel orbiting the field seeking to form up in the air. It is far better to get going on the first leg of the route immediately after takeoff, at reduced power settings. This naturally supposes that you took the time to form up on the runway first – if you didn't, you've already committed the formation to the first of many debilitating mistakes that will get it killed at the first whiff of combat.

Fig 74. Heading to the target area, an ad-hoc formation of Hurricane IIC's deploys in extended line abreast.

You might not believe this, but navigation is actually the hardest part of formation flying. You'll soon notice that merely sticking to an unerring speed, heading and rate of ascent is difficult enough. And at the first major turn, the ad-hoc formation is highly likely to start losing members who weren't paying attention. Therefore, announce upcoming turns well in advance, and fly as steadily as you can to facilitate formatting and observation.

WHAT RULES OF ENGAGEMENT?

The ROE protocol must be simple, and you must be adamant in enforcing it. A typical ROE for a high altitude sweep might include the following:

- The formation engages and disengages on MY order only.
- Hard deck is 10,000 ft (the altitude you are never to go below).
- Soft deck is 15,000 ft (go below at your peril, and immediately regain cruise altitude).
- The formation engages as a group in a single pass.
- Wingmen, stick to your leaders at all times!

19.1 FIGHTER SWEEP – RAMRODS AND RODEOS

The basic and most prevalent of fighter missions, the sweep is a free-ranging foray into enemy territory with the express intent of interdicting and disrupting enemy concentrations, thereby imposing your will unto the enemy. This is the classic air superiority mission, and it doesn't even have to net a bagful of kills to be successful. Success lies in direct and indirect ownership of the sky: even if the sweep is ignored or never even seen by the enemy it performs, at least hypothetically, a vital role in denying certain routes and chunks of airspace to the enemy for, and sometimes beyond, the duration of the flight. Should the enemy however direct resources to deal with it (resources which could be better employed elsewhere) even the mere tying up of enemy is productive - kills should be regarded as a bonus. Naturally, a sweep is all the more productive if the enemy is found and defeated, but that may not always be the priority – when the sweep precedes a bomber formation for instance, it is certainly enough to make sure that there are no enemy aircraft ready to pounce.

The key to a successful sweep is to maintain momentum over enemy territory and avoid becoming stationary at all cost, since the enemy is free to box you in and pile up the odds against your safe return. The objective remains to make your kills swiftly, to goad the enemy to commit against you, and to keep the flight going at all times.

Plan your sweep routes and altitudes well before setting out. Take into consideration where enemy is likely to be climbing and cruising, the inbound and outbound routes of friendly aircraft which are likely to attract enemy, the location of and distances to enemy airdromes etc. When the enemy is encountered defeat him quickly with only so many aircraft as are necessary, while the remainder keeps a lookout, and regain the forward momentum as soon as possible.

The sweep can be performed by as little as a single element but, naturally, the more the merrier. Under the best circumstances you have two flights of four deployed in line abreast. Following this formation, and ideally parallel to it, you have several other 8-ship formations scouring the skies. For, even though one brush of the broom might catch a few bandits, it can only cover so much sky and only for a limited time. To really own the sky you need to have consecutive and parallel flights covering enemy airspace at all times, and at all altitudes too.

The sweep is an important part of a strategic bombing mission (RAMROD). As such it precedes the bomber box with a separation of a few minutes, clearing a path for it as it were. Sweeps should ideally be conducted both in front and on the flanks of the bomber route and, yet more ideally, be backed up by a second and even third wave of fighter sweeps. In such a mission it is crucial to stay in radio contact with the bombers and to cruise at a reduced speed so as not to outstrip the plodding box entirely.

A RODEO is somewhat different. The historical Rodeo was a small bomber force heavily escorted by RAF Spitfires with the intent to entice the enemy to fight, or, in other words, a wolf in sheep's clothing. The British banked on the Germans to scramble their fighters, thence to be dispatched by the escorts. As it usually played out however, the wily enemy took the opportunity to make a few high-speed "gun and run" passes. The usefulness of the Rodeo is debatable: close escort is one of the most difficult missions imaginable and sheer numbers is as much a handicap as it may be a blessing.

19.2 COMBAT AIR PATROL - CAP

CAP is a point-defence (or attack support) mission tied to certain landmarks or map references. The mission is usually defence of an installation such an airfield, port or marshalling yard but all kinds of beneficiaries are possible, including friendly ground forces, convoys and trains. CAP can also be performed over an enemy airfield or piece of contested ground in order to facilitate the attack by other arms or squadrons.

CAP is the doorkeeper, the janitor who keeps the pesky enemy at bay. The mission may take place directly over the place in question, or along a more remote racetrack route in the most likely direction of enemy interference, although still within visual distance of the contested feature.

A CAP flight normally flies a predesignated track, usually in a box pattern with specific waypoints. Thus, if combat occurs, an engaged element can easily pick up the track again after concluding the battle – or call for help from hitherto unengaged parts of the patrol. The CAP flight should comprise enough pilots to deal with any threat the enemy might bring to bear lest it be overpowered and unable to fulfil its mission. The minimum is, as always, the fighting element, but a flight of four divided into two elements is of course preferable. If strong enemy pressure is expected, several flights of four may be required, and at varying altitudes to boot. It should also be borne in mind that a CAP flight requires timely reinforcement due to combat and the resulting loss of airspace control, ammunition and fuel expenditure, combat losses etc.

The most difficult part of flying the CAP pattern is to make positive identification between friend and foe. Many times the CAP must shed position and energy to ID bogies which often turn out to be friendlies passing through.

One of the most worthwhile and satisfying CAP missions is conducted at or in the direct vicinity of enemy airfields. Here, identification is rarely a problem except when the formation breaks up or when the elements are circling outside maximum identifiable range: everything can be assumed to be enemy. The destruction and disruption of enemy assets at the source is the most effective form of air superiority short of a tank on the runway – by engaging the enemy before he has the opportunity to spread out, while he's suffering under every conceivable disadvantage, he can be controlled and defeated with almost scientific precision.

This "Iron Hand" type mission requires excellent wide-area SA and a careful sense for when the force is about to overstay its welcome. While the force may appear and get to work with a sizeable element of surprise, this is quickly eroded in pace with the number of burning enemy aircraft. It generally doesn't take long for the enemy to wise up and contest your presence, even so, until that happens you'll get a lot of sweet shots on nonmanoeuvring outbound bandits. And when the jig's up, you can still claim ownership to the high ground and smack down anyone who comes to challenge it – until the numbers turn against you. On the other hand, if you bring sufficient numbers on your own you can be reasonably sure of extending your visit for hours on end (been there, done that!), or until such time as the enemy treats you with pertinent respect and amasses a likewise sizeable force to oppose you. This force is likely to arrive from the next nearest enemy airfield, so, if you can spare the bodies, detail an element or two to BARCAP that field too.

Key to efficient airfield suppression are, in no particular order: unfailing voice communications and radio discipline; target designation (i.e. the element next in turn to engage specify which of the many targets below they intend to engage and in which direction they're engaging); aircraft and pilots capable of producing one-pass kills; discipline in disengaging from failed attacks and the subsequent teamwork required to sandwich the enemy; the unbroken ownership of the high ground through maintaining at least an element at high altitude (top cover) at all times; discipline in formation disengagement.

There comes a time when the formation must break contact, and the call to disengage should ideally be made well before the situation gets out of hand. Here's one such encounter:

56th FG spent some time in the Metz area today, forming up over the target in wingpairs consisting bmbm/trimcz and hammered/salvo. Previously in the sortie we had been smacking down Metz pretty good, though at this juncture in the story some four-five 109s were observed taking off below. Rather than continue the orbit against a wide-awake enemy who would be evading and dragging us down in single melees, I decided for a change of plan: let the 109s take off while we disengage and head west, then we double back in two elements and bracket them. Thus we formed up in line abreast heading west at 4 km. Due south of Etain, we reversed and headed back, now with the elements wide apart. Too wide, as it transpired.

As trimcz and I passed Jarny (heading east) I spotted one bandit on our low 10 o'clock, then two more and following closely on their heels, two more. All headed west and climbing. I called the contact and winged over into the attack, trimcz following. I knew hammered and salvo would be some distance to the south and hoped they'd pick up the trail. The 109s scattered as I dove in. They were in two elements as well, the left in decent spread but the right in trail. The fifth contact was off to the right I believe. I went for the right pair in trail, but they broke off just in time. I kept on trucking, keeping my energy, knowing the 109s would take the bait. Trimcz worked through them as well, making them lose more energy in evasives. By now hammered and salvo were streaking in, but still out of visual.

I chandelled around to clear trimcz, seeing four 109s in pursuit. The left element broke as one, one high and one level. I passed them without as much as a glance though they were hardly more than 400 yards distant. The two closest bandits to trimcz were certain to get sixcalls and verily, they broke off just as I drew into range. I disregarded them as well and kept trucking again. Now trimcz reversed, just as hammered and salvo started to get in range. This the 109s had not reckoned with!

Hammered scored two 109s in this classic sandwich, alas, Salvo in a H75 was unable to keep pace and became saddled with three 109s (possibly another crew) near Etain while the rest of us proceeded to clear out the two remaining bandits. A swirling high-altitude battle commenced just east of Verdun, a slow motion ballet with infinitesimal opportunities at 7 km. Trimcz got a shot at one of the 109s who was seen diving out north trailing smoke – Trimcz followed him down and completed his demise. I was unaware of that at the time, knowing that maybe as many as three 109s were still around at 5-7 km – I wasn't fully aware of the fact that hammered had gunned down two already.

We eventually concluded that we were clear however, and proceeded home to land at Montfaucon and Reims. This was an awesome battle, a truly nerve-wracking experience of WWIIOL at its best.

The routine for whacking unaware outbound bandits is quite uncomplicated. Circling the enemy airfield at 3-4 km altitude (10-12,000 ft) it takes a very suspicious or wide-awake enemy to spot the formation. Most bandits aren't aware at all, to the contrary, most pilots expect their own field to be some sort of sanctuary. Error one. Since most bandits truck out on their lonesome (error two) and without rolling or essing to check their six (error three), and climb at max rate on a steady heading (error four), they present extremely simple targets for the high six bouncer. If the enemy doesn't climb radically or is in possession of a particularly fast aircraft the swooper does well in employing a low six bounce. Normally one lets the bandit take off and set out before swooping in, and if you're feeling particularly devious you may want to let them get out of visual range from the airfield proper so as to not tip off those still on the runway. Flame these easy targets in one fell swoop, break off perpendicularly to the enemy flight lane and regain orbit altitude before returning to the pattern.

One cardinal error most swooper-wannabees commit is to try and finish off enemy in the flight lane after a failed pass, or worse, fighting in the airfield AA. This ensures that other enemy flocks to the scene to overload the wannabee's SA and wrestle him down. If you botch your attack, break off and use team tactics to defeat the bandit. Drag and bag (i.e. sandwich) works great. And if the enemy should manage to wiggle out of control of the CAP force altogether and work up to co-altitude, it's a minor concession to sidestep until the area grows quiet again, then to return and set up the CAP one more time.

WIDE AREA DEFENSE

Organising the defense of an entire area, as opposed to setting up a basic CAP (the nature of which is a part rather than the whole) is a tough job that requires a fair bit of leadership and coordination between elements, particularly with regard to logistics (combat losses, fuel and bullets issues) and target handoff between elements. Figure 75 below outlines the basics of area defense at a hotly contested locale, comprising:

Low CAP, more or less directly overhead in a tight orbit of, say, a couple of miles between turns. This force should comprise sufficient numbers to take care of any threat that manages to filter through the outlying defense, at the least one flight of four and preferably a full squadron of 8-12 ships. This force should be on a rather loose leash and fly as wingpairs rather than as flights or squadrons, because in the Low CAP fight reaction time is at a premium, and bigger formations suffer longer reaction times. Make sure to designate a Rally Point for this flight in case pilots get separated for any reason, preferably some town or other conspicuous feature a few klicks away from the CAP objective where climbing and rejoining can be made with less risk of enemy intervention – this makes it easier for wingpairs to rejoin and avoids "false" bounces and panic alerts. Make no mistake: Low CAP is the last line of defense and relies mainly on overwhelming numbers and gaggle fighting to defeat the enemy: make sure your selected group understands what's expected of them.

High CAP, two to four elements making simple tracks along cardinal headings (E-W, N-S) or between specific landmarks in a box comfortably remote from the defensive objective. You will want to keep these tracks sufficiently distant in order to have at least a minute or two to identify and defeat inbound threats before they arrive over the objective. High CAP makes sure that Low CAP can go about their close defense without having to worry overmuch about badguys coming in hard on their six. In this case we assume that friendlies are out in force, so make sure to have at least one wingpair assigned to the "rear" of the area to guard against enemy trying to attack from the least expected direction. When performing High CAP it is good form to position two elements on parallel tracks, well in sight of each other, and have the elements perform cross-turns at the waypoints. The track length, if perpendicular to the main enemy ingress route as in figure 75, should be approximately 2-3 flight minutes long: from the cross-turn, you have 1 minute to observe the track in your forward quarter, 1 minute to cross over it and 1 minute to continue past it to the opposite waypoint where you again cross-turn back on track. It's tedious and sometimes downright boring but it pays off. Big time. If you do your job well enough, and are suitably positioned away from the objective with another element as backup, Low CAP will have a very easy time of it, to the point of complaining of the lack of business.

BARCAP, requires two or more elements patrolling astride the main enemy ingress route. Their task is of course to defeat the "lone truckers" but also to own the highest altitudes by pushing down the enemy below High CAP altitude, thereby making their job all the more easier. Whatever threat they cannot easily handle without sacrificing position and/or mission integrity, they report to the main force. In this way, BARCAP paves the way for High CAP who paves the way for Low CAP, who makes sure the grunts are happy.

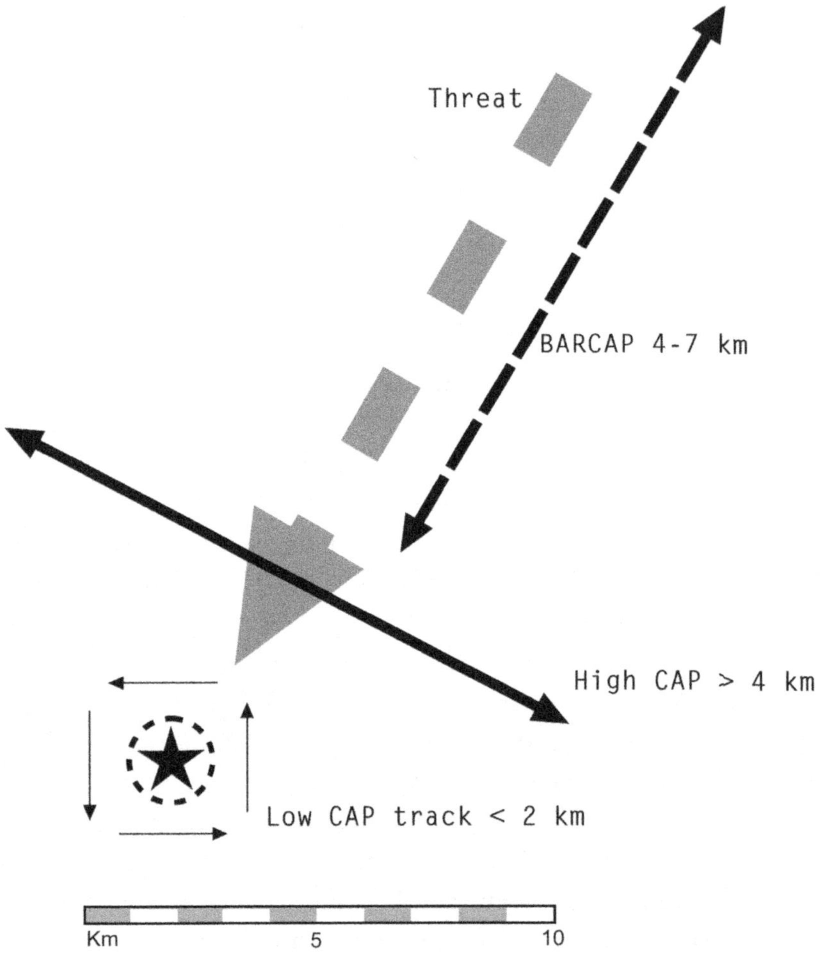

Fig 75. The Defensive Battle Plan.

19.3 BARRIER COMBAT AIR PATROL - BARCAP

BARCAP is a combination fighter sweep and remote CAP used against known and suspected enemy points of ingress to a contested locality. This is generally an ad-hoc mission, but a major planned offensive benefits greatly from the advance planning and staffing of several BARCAP locations insofar as it blocks the enemy from entering the combat area proper. Barrier CAP buys time for the main force, delays and disrupts the enemy and, in the best case, destroys him altogether. While the BARCAP might not be able to stop every enemy plane from getting through it certainly plays a very important role in reducing the enemy threat. Merely denying the enemy access to certain altitudes or vectors may be sufficient.

The BARCAP is, like the CAP flight, flown as a box or line pattern between specific waypoints at specific altitudes. The BARCAP team has a somewhat easier time in correctly identifying bogies since the bulk of outbound aircraft are likely to be enemy. Here, extra care must be taken to avoid extended combat: the closer the BARCAP is positioned to the enemy airdrome the greater the risk of it being overpowered.

Personally, I prefer flying BARCAP tracks perpendicular to and crossing over the expected enemy route. This allows me to sight the enemy over the wings, whereas a parallel track may hide the bandit below the bonnet and below my tail. The perpendicular track also facilitates better timing of the bounce in my opinion, but this is of course subject to personal preference.

19.4 CLOSE AIR SUPPORT - CAS

The crunchies love air support when it's on their side, and hate it when it's used against them. CAS (Close Air Support) flights should have their own escort (or draw benefit from CAP and BARCAP missions already in progress) since it's asking too much of the CAS flight to perform two missions in one.

That said, the CAS flight needs to have a standard response for the advent of enemy interception, or enemy presence at the target. If attacked enroute to the target, it is advisable to jettison any external ordnance and prepare to receive the enemy as in any other fighter-vs-fighter situation. If the enemy is encountered co-altitude or better (from your point of view), simply blow past or ignore them until you've delivered your goods.

Always attack the ground feature on a homeward heading in order to avoid energy-consuming turns in the presence of an angry opposition. Make one pass and go home, unless of course the feature is totally undefended. In such a case it can be highly entertaining to strafe enemy troops and transports, antitank guns and so forth. A measure of caution is advisable however as your wide-area SA degrades quickly when you're on the weeds. Make a habit of checking your six at all times, even when strafing, and make the occasional sweep around the area at somewhat higher altitude to clear it from airborne threats. The most odious threat is naturally poised by enemy AA (apart from trees and terra firma). If such obstacles to your strafing enjoyment is encountered, make certain to attack them with more than one aircraft: the AA usually targets the closest threat and is then vulnerable to a follow-up attack on the heels of the first. Should you, God forbid, be all alone in the area, either stay well away from AA, cast your eggs (if any) on them or attack them creatively: if your first attack was high, exit in the weeds and return in the weeds from another direction to strafe them dead.

Communication with friendly ground forces is auspicious, particularly as you will not have much chance of noting the difference between friend and foe. Attacking your own forces is not appreciated by anyone except the enemy. If possible, have the crunchies mark the target with smoke or tracer. If you make repeat sorties to the same target, be sure to vary your ingress direction and altitude to make yourself less predictable.

Keep your speed up when performing CAS! The AA gunners positively love slow and predictable targets while drawing lead on fast fighters who expose themselves only for a few fractions of a second is nigh impossible. Don't listen to pilots who say that they slow down and throw out the boards (flaps) when strafing, they are clearly delusional. You will want to fly as fast as you possibly can, and remain as unpredictable as possible.

When you're doing CAS you are well served by having at least an element out on BARCAP reasonably high in the most likely enemy ingress vector. They may not be able to deflect every bandit but they should be able to give timely warning of impending party-crashers. That too is a reason to remain at high speed: you can't very well perform adequate guns defence when low and slow. Jam that throttle forward!

19.5 ARMED RECONNAISSANCE - RHUBARB

If you have nothing better to do, and fancy a lingering thrill, consider going out on a low-altitude armed recon run. Like all others, armed recon missions should be well prepared and briefed since the likelihood of things going catastrophically wrong is much higher the closer you get to the ground!

Route planning and memorizing of prominent ground features are essential - you will not have much time or opportunity for detailed map studies in the cockpit.

Armed recon missions are generally flown at minimum altitude and at best sustained speed, although it is eminently possible to go in at 5-6000 feet. Be advised that the higher you go, the less you see, and the better you are seen. Staying down in the weeds is a good bet, but you will only get very brief firing opportunities.

Should you happen unto a juicy target such as an undefended convoy or troop concentration, why, feel free to zoom, stay a while and go wild before exiting the stage, again in the weeds.

Another way to maximize your firing opportunities is to have a point flight a minute or so up front to report on targets of opportunity, allowing subsequent flights to adjust their headings for a clear run.

Avoid known AAA installations and only make single passes on high-value localities such as enemy airfields, for your own good. If pounced upon by enemy enjoying a higher energy state, use separation between flights and weave them to death. That, short of a miracle, is your best bet of seeing home turf. Avoid becoming stationary at all cost. Flights should maintain line abreast formation unless the territory forces you to employ line astern. This takes a lot of effort and discipline, but should be second nature for the veteran pilots.

19.6 BOMBER INTERCEPTION

This mission is pretty straight-forward: follow directions from ground- or airborne observers, gain as much altitude as needed depending on the enemy escort situation, and set up a good clean run against the bombers. It is auspicious to manoeuvre yourself into a position with the enemy escort (if any) in line with the bombers so that you may blow by them with a minimum-separation pass either before or after you blast the bombers. If you have the energy advantage, extend away from the escorts and set up another high-speed pass. Rinse, repeat.

Never ever climb up toward the bombers within escort range - it's certain to ventilate you in any number of undesired places. Whether you target the bombers or the escorts first is wholly dependent on the situation: performance, relative energy states and estimated time to target should be the primary considerations. If you have plenty of time, get some altitude and clean out the escorts. If you're in a hurry, target the bombers. If there are no escorts, woe betide the bombers! The actual attack has as many options and gambits as combat in general. Against unprotected bombers, as described in the ACM section, a high sides attack is generally the most profitable. This exposes the enemy the most while presenting his gunners with a particularly tricky solution.

In the best of scenarios you occupy the high ground already and may choose your type of attack. In the second best scenario you engage the bombers either head-on or from their forward quarter, and have energy enough to swing around and go for seconds (beware the escorts however!). The most common situation however is a stern chase. It is here that you must fight the urge to motor up on their low or level six o'clock for what seems a lazy low-closure gunning session. In this case you're the most vulnerable not only to tail gunners, belly gunners and dorsal gunners but to the escorts as well. Your situational awareness will be in the pits as you hesitate to err from your pursuit heading; you will be low, slow and utterly predictable.

If you are indeed coming in from astern, take a vector that will bring you alongside the bombers at slightly more than a full turn radius away. Climb beside them and when you're at their ten or two o'clock high, wing over into a high sides attack that brings their full planform into your sights. Completing the attack, dodge whatever defensive fire is brought to bear and regain your high side position on the opposite flank. Repeat until they're done for. If there are escorts present, initiate the attack with an element or two to draw their attention while holding your main force back for the real attack. Only rarely will you have enough fighters around to completely annihilate the escorts: your best bet is to draw them away from the bombers with but a small part of your force.

19.7 BOMBER ESCORT

Bomber escort is a science unto itself, and we will only explore it briefly here. Escort is arguably one of the most difficult missions you can sign up for, but also one of the potentially most rewarding. A thoroughly planned and staffed bomber mission include several types of escort and supporting missions, fighter sweep being one of them. The actual escort missions are:

REMOTE ESCORT

While the fighter sweep cuts a deep swathe in enemy territory and poses a threat to enemy climb out vectors, the remote escort's chief mission is to detect and deflect enemy interceptors directed against the bomber formation. This is best performed at a distance of a few minutes ahead and out on the flanks of the bomber box - not too close, and not to far. Knowledge of the bombers' route and progress is paramount: the flight leaders should be in communication with the bombers at all times in order to position themselves correctly and to defeat inbound enemy.

The remote escort's mission is NOT to seek the absolute destruction of any enemy encountered unless of course such a happy opportunity presents itself. Rather, its mission is to maintain the integrity of the bomber box by deflecting enemy interceptors and deny them access to position from which they can pose a threat to the bombers. A simple bounce or even a moderately threatening gesture may be enough to force the enemy to abandon their mission and/or cede altitude. As always, however, circumstances dictate whether the remote escort should maintain position relative to the bombers or go balls out against the enemy: if the bombers are at risk, anything and everything should be done to reduce or neutralise the threat.

CLOSE ESCORT

While sweeps and remote escorts (if any!) may be able to deal with some of the enemy, the bomber crews very much appreciate a close escort and rely on them to sort out such enemy as leak through the screen. Close escort should place themselves ahead, well above and out on the flanks at the box's 10 and 2 o'clock, with the bombers comfortably within visual distance. Be advised not to fly too close however, as you will need some time and distance to ward off high-speed attacks. Don't ever fly so close to the bombers as to be a part of their formation – there is absolutely nothing you can do against enemy interceptors if you mingle with the bombers. Maintaining these flanking stations is imperative for several reasons:

- Ease of identification and ease of leadership. Any bogey spotted outside of the escort stations can be assumed to be hostile and the appropriate measures leveraged against him.

- o Reduced SA load. It is far easier to keep track of the bomber box, and of bogeys and squadron elements, if the bomber box is kept continuously in sight over the wing, in comparison to a flight profile where you wander all over the sky.
- o Reduced fuel consumption. Keeping pace with the bombers allows you to cut back on boost and RPM, thus increasing your range and combat effectiveness.

Always attack with the enemy placed between yourself and the box, and break off as soon as you see the enemy abandon the attack. Your mission is to protect the bombers, not to become locked in individual combat well behind and below the box. Keeping a healthy distance to the bombers is vital: if you fly too close you're essentially part of the same target group and will be largely forced into frontal attacks against engaging enemy – or forced to chase the enemy away from the bombers after they've successfully attacked the bombers already. Thus, you need to maintain a distance to the bombers that allow you to spot, identify and react to enemy fighters. You need time to communicate the threat to the other escorting elements, time to decide on the best course of action, time to wing over into the attack, time to close the distance, time and distance to defeat the enemy before they get to guns range.

In the following picture, the 10 o'clock high station escort spots a low closure threat in the bombers' rear hemisphere. The escort turns gently back and executes a standard high rear attack against the threat. If the enemy is any good he'll break off his attack as soon as the escort makes a threatening move – in that case the escort should resume his station while keeping an eye on the bandits. If the enemy is fixated on the bombers he'll be set up for swift annihilation in a single attack, whereupon the escorts are free to regain position. On no account should the escorts sacrifice their energy advantage – which is considerable even if they're motoring along at reduced power – and pursue the enemy away from the bombers. Merely bouncing them away or scaring them sufficiently to abandon the attack is more than enough. However, care should be taken to fully engage high-energy threats until destroyed or neutralised! Thus, if a low-energy threat is judged to remain in the fight and eventually become a high-energy threat, defeat him while you still hold a significant energy advantage. If not, things will become all the more trickier later on.

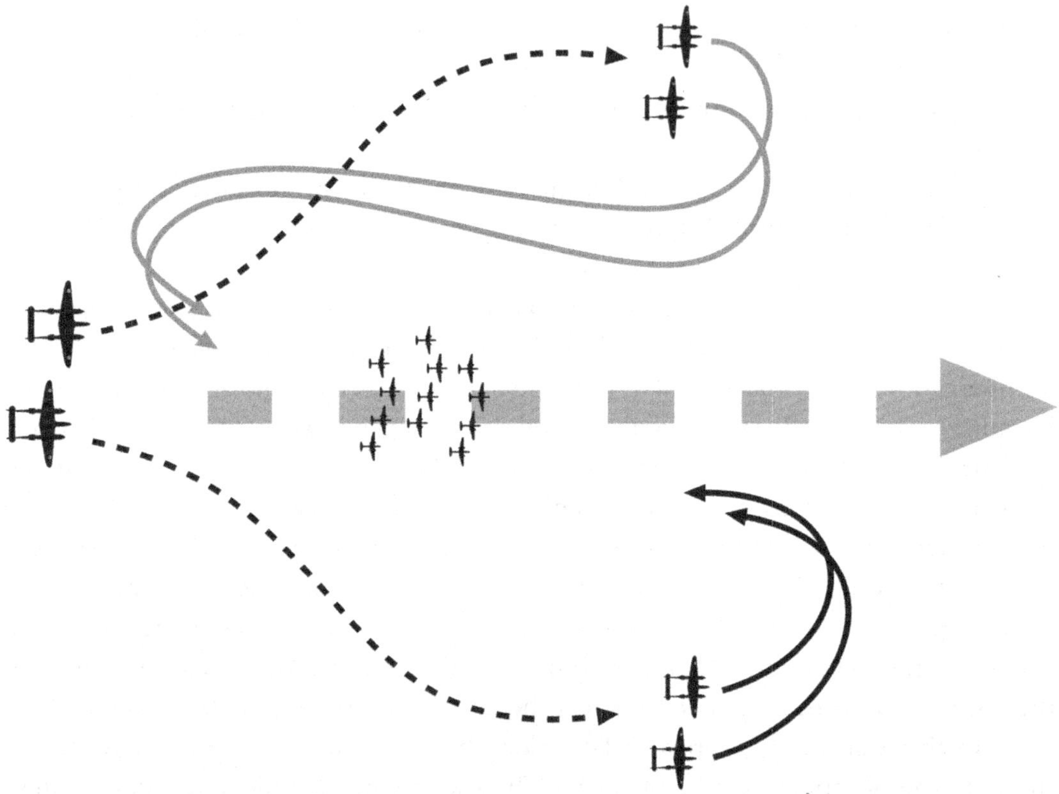

Fig 76. Escort stations and reaction to enemy fighters. In this general schematic the 2 o'clock pair takes on a threat approaching the bombers from head on. The 10 o'clock pair takes on another threat approaching the bombers from astern. The high six reserve flight serves to cover either flanking pair as required and assumes their station when either pair engages.

If your numbers are low, break up the squadron into elements so that you cover both flanks. If there are several escorting squadrons, assign one sector to each. It is also good form to have a small reserve high on the box's rear quarter, if you can spare the manpower.

The really critical part of the escort job is when the escorts themselves are singled out for attack, and found at a disadvantage. Although you're morally bound to protect the bombers you should never hesitate to save your own skin - if you die there's one less escort anyway so you might as well save your country the expense and get the hell out of Dodge! With any luck and a measure of cooperation, your wingman might even save your bacon and allow you to rejoin the mission.

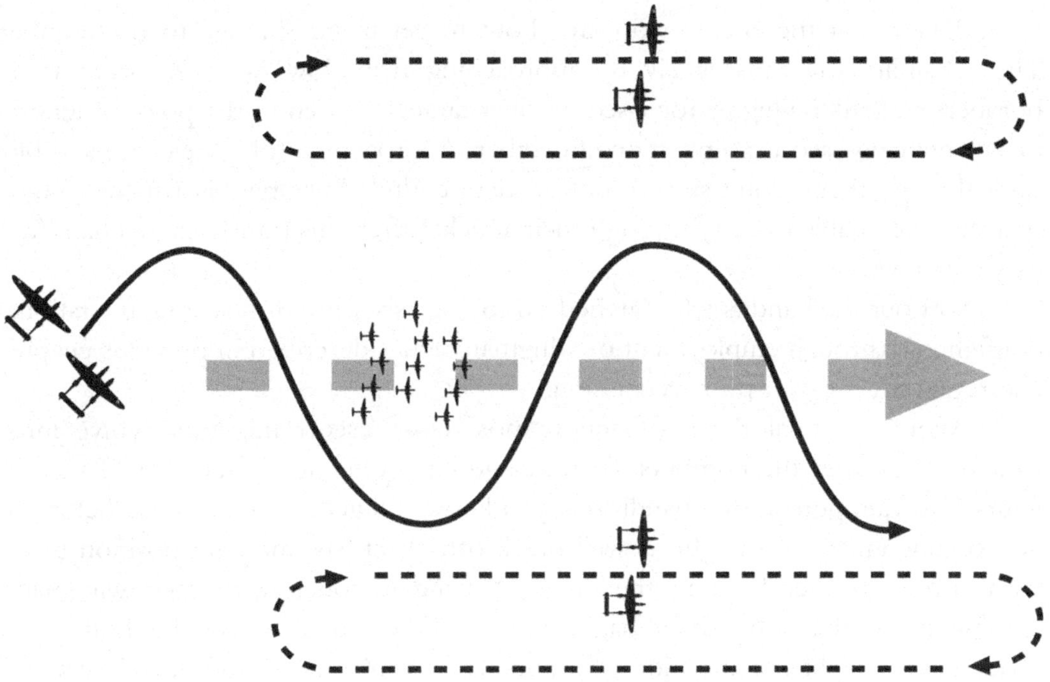

Fig 77. Managing escort rate of advance relative to the bomber box is crucial. During climbout and cruise in friendly airspace the escorts may cut back on throttle and remain largely constant in their allotted positions. When crossing into enemy airspace however, escorts need to be operating at a higher readiness and thus a better turn of speed is required. Here can be seen how the flanking parties are maintaining their general positions by flying parallel tracks, using either cross-turns or in-place turns to stay in line abreast. The high six party is better off flying 90 degree doglegs in a weaving path behind the bombers, using tactical turns to stay in line abreast.

At or shortly after the target is where close escort earns their pay. Here the bombers will be subjected to a mass of attacks from all quarters, taxing the escorts to the max. Whereas a fraction of attacks will be delivered from a position of advantage the bulk of interceptors will be attacking or preparing to attack from low astern of the bomber box. Now the escorts must perform a split second triage and prioritize among threats. As always the high-energy bandits should be your first concern.

Attack the enemy who are in the most threatening position first, taking care not to pursue them futilely so that other low-priority threats are given time to become high-priority threats. If you cannot kill them outright, forcing them to abandon comfortable positions for subsequent attacks is about the best you can hope for. Threaten them only so much that they must spend time and space in preparation for subsequent attacks, all the while keeping your energy up in the direction of the bomber heading.

Attack first the enemy who are about to get in guns range to the bombers. These bandits will most likely be approaching from low or level astern to the bombers and will be highly focussed on their targets – often to the point of ignoring or not noticing escort intervention altogether. Approach with high closure and blow right through them, don't stop to defeat them entirely but only blast them so much that they are unable to carry through their attack. Often this translates to a handful of easy kills anyhow.

Sort out the bandits who latched on to your six without resorting to stationary dogfighting, through employment of wingman tactics described in previous chapters, and regain SA for your primary mission.

After the initial flurry of action most close escort missions evolve into a running fight after the bombers, friendly and enemy fighters interlaced. If you find yourself in vain pursuit of a bandit that you know will get to the bombers before you get to guns range, you might as well break off, check six and gain position to deal with anyone further back in the queue. Staying in touch with your wingman is paramount in this long drag&bag situation. You can also use the bombers as "wingmen" to split the attention of a pursuer: if you're alone and have a bandit on your six, vector some 30-45 degrees away from the bombers to force the enemy to choose between futilely chasing you or productively pursue the bombers. Once he breaks off (if he breaks off!) toward the bombers you can swing back into him and defeat his pursuit. Be wary of the vectors involved and the timing of this manoeuvre! It is not uncommon to end up way behind the bandit and thus being without a say in the proceedings. The wise escort plots his intercept so as to keep the bandit off his wing while flying "pipper on" the bombers – in this manner you fly an "automatic" lead pursuit on the badguy and can fairly easily cut in behind him as long as he stays focused on the bombers.

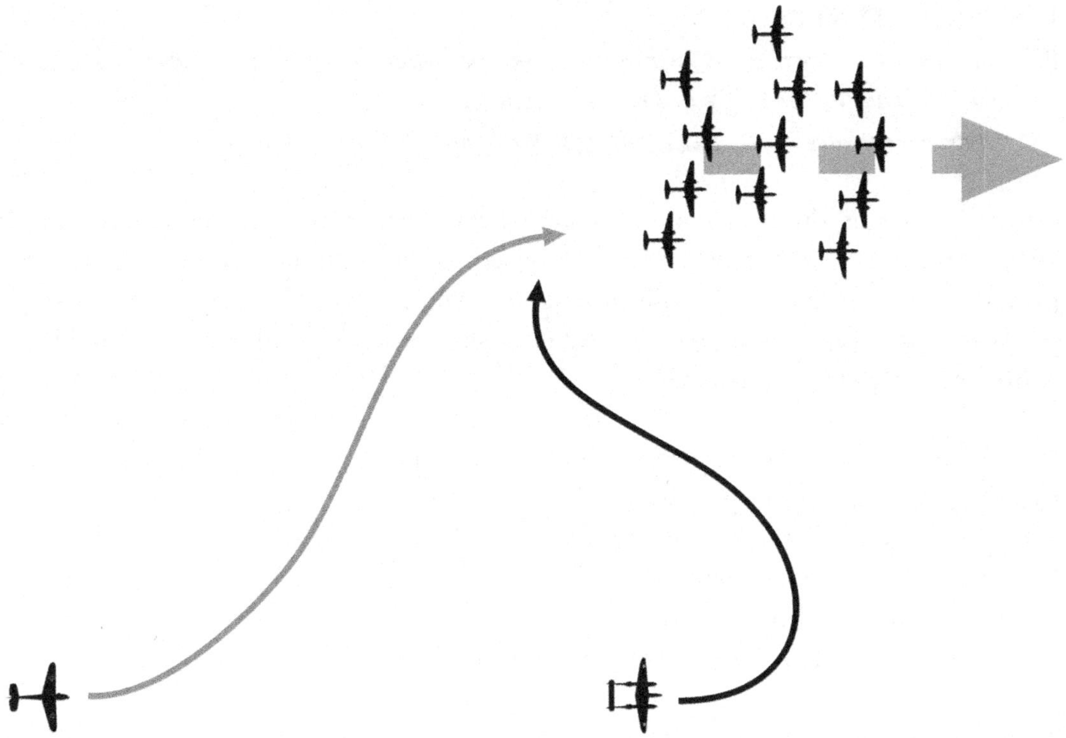

Fig 78. Tactical use of bomber box to shift pursuing enemy's interest from outnumbered escort (not to scale).

RECEPTION ESCORT

Since close escorts have a tendency to become stripped from their charges, a full-blown raid should include a friendly reception party. This mission launches well behind the bombers and rendez-vouses with them at a previously arranged spot and time, generally located three-quarters of the distance to the target (provided the fighters have sufficient range). Communication with the bombers is recommended so that the reception can position itself to deal with any remaining interceptors. From there on the reception escort continues as close escort. If contested, the bombers are certain to have stragglers with ailing engines, in which case you'd do well by assigning a pair of fighters to each cripple.

19.8 GROUND ATTACK

We have the enemy surrounded. We are dug in and have overwhelming numbers. But enemy airpower is mauling us badly. We will have to withdraw.

Japanese infantry commander reporting to headquarters, Burma, WWII.

Ground attack is different from Close Air Support insofar as the ground attack mission is more a one-off mission as opposed to the more persistent CAS mission. Both, however, require the establishment and maintenance of air superiority – this is ALWAYS the case, as stated by Air Marshal Tedder in the 1943 Casablanca conference, in point 5 of his dicta:

The fighter plane is the basic weapon of an air force. It should be used for the following missions in this priority:

a. Fighter sweeps to clear the enemy out of the sky.

b. Escort for light and medium bombers.

c. Interception of enemy aircraft.

d. As a fighter bomber to provide CAS for ground forces.

The ground attack mission, quite as any other mission, requires mass to be effective: the discerning party sorties in strength and attacks in a disciplined and concentrated manner against an identified feature or target. Simply winging along and attacking any old target of opportunity on your lonesome is NOT a proper way of doing it, mainly because your singleton performance is less than likely to produce adequate results and because it only takes one enemy fighter to render the whole endeavour futile. Go in force!

Leadership, control and discipline is what makes the ground attack effective, and with that in mind several factors become apparent as being partial to mission success:

- Intelligence: Type and location of target, type and location of AA, overall air situation, time available.
- Loadout and mission profile: Is it a guns only attack? Do we bring bombs? Rockets? Ingress direction; cruise altitudes etc.
- Coordination with ground observers and/or ground controllers: exact location of target relative to conspicuous ground features; is smoke available to mark the target?
- Air superiority: who has it; coordination with fighter cover; time allotted for creating a window of opportunity?
- AA suppression: how to minimize threats from ground fire.

- o Ordnance delivery: area attack; attack individual targets; how many passes are allowed etc.
- o Rally point and egress: where to recover and/or reform etc.

From this simple template it should be obvious that ground attack is something more than a simple "dive in and chuck that egg" routine: it should be conducted with as much preparation and teamwork as anything else to ensure success. Let's look at how a sample mission might be conducted from a "best practice" perspective.

You have been alerted to the presence of a strong enemy armoured force approaching a key point. The call goes out to the air force to reduce the threat. The enemy air force is out in strength so you need to coordinate your arrival with the air superiority fighters – and you need to get moving PDQ lest the armoured column arrives intact. You assemble your squadron, plus a number of freelance fighters of unknown capability, on the runway of the nearest available airfield. Sitreps keep coming in on all frequencies – the armour column is closing fast and enemy air is highly active. You have six bomb-carriers, and the support of an unknown number of friendly air superiority fighters at the objective. It is reasonable to assume that you will encounter little in the way of AA, and that you will only have the opportunity to make a single attack without enemy intervention.

Time is of the essence so you get going with a gentle climb to altitude, keeping your speed up – though only so fast as to keep the strike force together. Wingmen are assigned on the fly. You get in touch with the fighters on the scene, announcing your TOT (time on target) and requesting maximum support for your attack. A pair of air superiority fighters attach themselves to you during the ingress, flying herd on the flanks of your little formation. You swing round the objective area at altitude to gain a feel for the overall situation: the fight appears to be low and not directly over the road in question. Now, where are those tanks? Somewhere southeast of the town is the last known location. You need better confirmation than that, and you're loath to go down and blow your energy and surprise with a bunch of enemy fighters around. As you curve around onto the enemy ingress direction the necessary spotrep is given: "30+ tanks, 500 meters east of the wood straddling the E-W road east of town X, all tanks on the road in column." That's all you need to know.

You issue your orders quickly: "Fighters, take up station over town X to clear our egress. We will attack from east to west. Squadron, follow me in trail! One pass!" You see the fighters peel off to take up station, widen your turn to give them some time and tip your wing to line up with the E-W road. You're attacking from 3000 meters so you will want to do it at a pretty low angle: 20-30 degrees. There's the road. Angle looking good. Your squadron is falling into trail behind you, nervously checking six and flipping their safeties off in preparation for the attack.

In in in! Winging over and diving down you roll the trim wheel forward and pick up the road. There it is, and down there somewhere is the armour column. Passing through 1500 meters you spot the telltale dust clouds and then the great metal mass – here goes! Aim for the head of the column, hope to nail the leader. You're hurtling down at 550 km/h, your aim is true, you pickle both bombs as the gunsight passes over the lead tank, trim the nose up and scram west, going lower and lower until you buzz past below the treetops. Dull thuds behind you, pillars of dirty black smoke rising. Past the town you quarter-roll and check six: five planes behind you, no enemy as far as you can see. Your covering fighters report: clear! You reduce throttle to facilitate formation and continue home on the deck, your pilots shaking out in the familiar wide line abreast. Done!

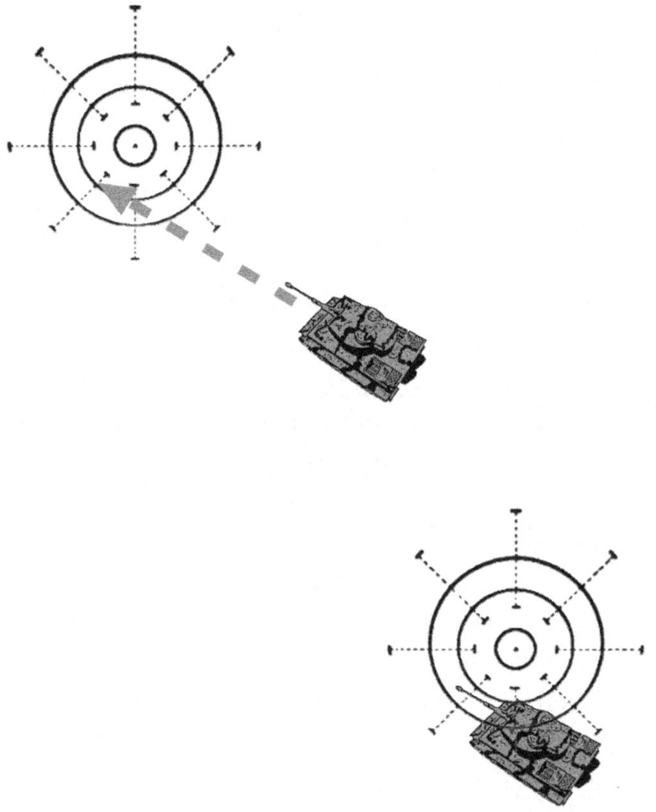

Fig 79. If the target is moving, aim slightly high, to allow for gravity, at the point where the target will be when your bomb arrives. If the target is stationary, aim just above.

The mission just described takes little more time than had you gone on your lonesome to the target. Not only does your survival rate skyrocket, you make a wonderful appearance and have fun too. You return safely to base, load up fresh bombs and assemble on the runway for a second helping of whup-butt. Stragglers delay your takeoff by a few minutes, and the climb to 3000 meters may seem excessive to pilots otherwise accustomed to dodging trees, however, the results (and the experience) should make it all worthwhile. Try it why don't you!

When attacking a locale defended by AA, it makes a lot of sense to draw their attention and (ineffective) fire with a couple of "stooges" while attacking them with planes held higher in reserve. Once the AA is down, go about your business as usual.

PART VI: COMMUNITY

"If it wasn't for the community I'd have quit the game years ago"

The difference between entertaining yourself with a single-player flight simulation facing artificial intelligence enemies and that of a massively multiplayer online roleplaying game lies not merely in the quality of the opposition but in the interaction with friends and foes alike. The MMORPG is a world in and of itself whereas the stand-alone simulator is just… a game. The MMORPG world is a dynamic, chaotic, universe where anything can and does happen. It's full of surprises and whacky encounters, dependencies and teamwork, grand operations, fog of war and friction. In a stand-alone sim you can be pretty much sure that things will happen in a certain way; that you will have a preset amount of time and support for your mission against a known number of enemies. It just doesn't happen like that in the MMORPG.

Take a simple fighter sweep for instance. In a stand-alone game you may be leading a flight of four ships but not really feeling "in command" or feeling any particular kinship with your flight members. They are no more than a couple of lines of computer code, set to act in a certain way. You cannot joke with them or tell them to behave in a particular manner, you can't ask them whether they have your six covered or ask them to drag a bandit in a certain way. They're just drones and you're just manipulating computer code. Fun? Not in my humble opinion.

In the MMORPG your flight members are real people. They often fail in keeping formation, fail in scanning diligence, they often err in navigation and make catastrophic errors of judgement. On the other hand, they may follow you slavishly and spot bandits way before you do, they will surprise you with acts of sudden brilliance and comments that will leave you writhing on the floor. When they're dragging a bandit you can direct them so as to clear them, you can devise ruses and play gambits, totally surprise the enemy and on occasion, get more than you bargained for or ever thought possible. No mission is ever the same. You will experience anticipation, pride, fear, anger, cowardice, bluster, joy, glee and next to every conceivable facet of human emotion in this game. Your sphincter will pucker, your cheeks will blush, your palms and armpits will break a sweat, your heart will pound and you will be jumping up and down for joy… or frustration. All of that and more is part of the game, of the game world.

Then suddenly I spot a shape in my slightly high 2 o'clock. Thinking it was Mick who had slid ahead I was just about to say something and power up to adjust formation when I saw that it was a 109, not Mick. I nearly fell off the chair! He was flying parallel no more than 250 yards out, slightly high and perfectly true. LOL! "109 my two o'clock, he hasn't seen us!". I went through the gate and slid to right, putting myself directly astern and somewhat low at perfect range. I fingered the tit, then hesitated. This one's Mick's! "Mick! You take him! See him? You have a shot?"

*"Yar, I'm behind you, I'll give him a plink from long range to wake him up" came Mick's deadpan reply. I was laughing so hard I nearly shot him down myself but restrained myself diligently hehehe. I saw Mick's tracer pounding the 109 dead astern, bits and pieces flying off, producing a trail of smoke. The 109 shuddered somewhat and slowly went into a steepening dive, opening up distance. I followed suit, going to continuous RPM to increase speed in the dive. And then *poof* he was gone, evidently a pilot kill. We recovered over Neuvillette and motored home, laughing lustily all the way!*

In the dynamic and persistent game world real people form friendships and forge alliances, devise strategies and tactics, agree on joint operations and engage in diplomacy, out-of-game communication and community building that transcends the suspicious hostility between opposing camps. The community arranges real-life meetings and even conventions, the community argues, bickers and taunts, it creates artwork, poetry, events and programs. Without the community the game is nothing. I have met many fellow virtual pilots in real life, both locally and internationally, for simple get-togethers and for playing sessions, for discussions and lectures on air combat. In some cases friendships have formed that exceed what one might reasonably expect from an online game, which is pretty fantastic and inspiring.

It is hard to engage in a MMORPG without making friends, without starting to recognize players or without making a "name" for yourself. Sure, some folks are more talkative or conspicuous than others, some keep a low profile and a lot of players never become community entities at all. But if you stick around for a couple of weeks and frequent the discussion forums and, better yet, become a part of a regular formation, you're highly likely to find new friends and new dimensions to the gaming experience. You're likely to have the time of your life. So much in fact that it becomes a very real part of your "other" life – for better or worse!

CHAPTER 20. SQUADRON LIFE

After you've mastered the flying and fighting part of the online game you will begin to appreciate the benefits of teamwork and community so much that you will want to be part of a formed unit. You've been around for a while now and know that there are squads of various makeup and composition in the game: funseekers, grognards, posturers, men of leisure and men of intense devotion to their steads and chosen allegiance. Take a pick, and if you're not perfectly happy with your choice leave and try another.

Leading a squadron is very much similar to leading a small firm or enterprise. As the commanding officer you get to define the formation in every respect from name and concept to mission types and guiding principles. The decision to form a squadron of your own is, I suppose, borne out of disappointment with how things are done in your present squadron or, in case it's your first shot at leading a squad, an urge to assure yourself of teamwork in order to excel or, if you will, increasing and sharing your enjoyment. I guess motives vary: mine certainly did.

As commanding officer of an outfit you accept a certain responsibility: to keep the unit in being; to gather your squadron and provide leadership in the game and outside of it; to expand your ranks and to provide enjoyment. You are the focal point of the squad's activities, you set its agenda and modus operandi, you decide how and when to play the game. Such an endeavour requires your full attention: you need to provide adequate communication channels, define the standards of membership, train new pilots and educate the unit in unit lore and fighter tactics. You can take this to any extreme: some squads have a stringent code of conduct and participation requirements, others are happy to remain a loose confederation of brothers in arms who get together whenever they feel like it.

Squadron life has a strong social aspect, as one might expect. It's hard to run a squad without sharing experiences, without learning where people are from and what they do for a living, without sharing a common ground with regards to the game and how to relate to it.

There are no rules for how to run a squadron: you make your own. Personally I subscribe to the notion that flying becomes a whole lot more fun if you attempt to do it with a measure of excellence in execution and style, meaning, a thing worth doing is worth doing well. That leads implicitly to doing things "properly" with regard to tactics and mission profiles, rather than a slapdash or "gamey" approach to the challenges of online flying.

There is no doubt in my mind that if you do things properly – such as is taught and expected in the real world – you will do a whole lot better in the game and consequently enjoy it so much more. Consider also that when you make an effort to fly in a realistic and efficient manner (think formations here) you also provide your opponents and onlookers with enjoyment. Surely it's a lot more fun to encounter a full-size formation that acts in a professional manner than a slew of unorganised singletons. Achieving such a level of teamwork takes effort, an effort which not everyone is prepared to accept, an effort which, when it's flaunted, makes all the difference between recreation and experience.

Morale is an important factor in the success or failure of your group as a whole. If your missions repeatedly end with the utter destruction of your squaddies, if objectives are but rarely met, if accusations are bandied back and forth, if no one communicates… then you have a morale problem that needs solving ASAP. Achieving mission success (however you define it) is imperative for keeping your guys happy: take them out, fly in style, down enemy by the bushel and bring them all back safely and you're certain to have a house full of happy campers. Building good morale, or esprit de corps if you will, is an ongoing task that will leave you stumped from time to time, as it's sure to plummet if left unattended. Running a weekly mission and keeping track of participation is a good start. Next, running survivable missions with clearly delineated objectives and responsibilities is another. There's more: writing after action reports; giving praise and encouragement where applicable; handing out virtual rewards for accomplishments and maintaining a visible and benevolent leadership profile – all of this and more serve to keep your blokes happy, cooperating and communicating.

Nothing succeeds like success. When your formation delivers or otherwise excels, boast and revel in your success! The highest mark of success is when your crowd is mentioned with respect and awe by the community, and when your inbox starts filling up with requests to join. Be aware, however, that morale is fickle and needs constant attention. A good leader delegates without relinquishing control, he nips conflicts in the bud and takes the faintest whiff of discontent as a cue to step up his community-building activities.

20.1 TRAINING

By definition, a squadron operates as a unit and must practice to become proficient in its chosen role. Merely taking off together and possibly sharing a voice communications channel is far from enough: you as commanding officer are responsible for the performance of your outfit and you'd be a sorry officer if you didn't endeavour to improve on the performance of your squadron. This book contains enough topics to keep you occupied for years to come <grin>, the question is rather how do you train and where to begin?

Clearly you must start by assessing the ability and potential of your squadron members: can they fly at all; what's their SA like; do they know how to fly formation; are they familiar with team tactics; do they know how to execute standard missions;

do they return to base on a regular basis etc. Once you've ascertained the training needs you can go on and create a simple training programme.

A lot of training can be conducted individually, offline. Take off, landings, advanced manoeuvres, strafing, performance testing etc is easy to practice on your own. One thing in particular that I like to recommend is flying (offline) while looking backwards only, including takeoff and landing. This simple exercise helps the pilot get to know his plane, his speed and attitude, by feel rather than by looking out front and watching the horizon, and it is also a good primer for flying "backwards" when keeping visual contact with a bandit in a scissors or barrel roll.

The things I normally do when faced with a new pilot are:

- o confirm that the pilot has his controllers set up appropriately
- o confirm that the pilot can taxi and takeoff without messing up
- o confirm his SA level by flying "follow the leader" and simple mock combat
- o confirm his ability to format in line abreast and to navigate using tactical turns
- o confirm that the pilot can perform guns defence and basic evasives to stay alive

The real mark of squadron proficiency, in my opinion of course, is its return ratio. Can the formation remain combat effective throughout a mission, does it return to base intact and what sort of damage did it deal per loss (if any). If you realise that your squad isn't performing too well along any of these criteria then you have some serious training to do!

It seems to me that a lot of groups are reluctant to train at all: they are content with going out to do combat as best they know how and that's it. That's too bad, for every mission offers a whole host of training opportunities, if realised as such: formation takeoff; formation ascent and cruise; navigation; change of formation; engaging by elements; employing top cover; voice comms practice; gunnery; evasives; drag&bag etc.

Critique your missions! Write after action reports and discuss what went wrong and what went well. Talking about situations and engagements is one of the best ways of learning and imparting knowledge, and in this racket you can't know too much.

When running a mission, make sure to hook up the rookies as wingmen to the accomplished pilots. Putting two greenhorns in an element of their own is likely to result in both their losses without offering a learning experience in turn. If a guy is notoriously bad at formation flying, format on *him* to show how it's done rather than keep goading him over the blower.

There will be certain missions that your unit likes more or is particularly good at. Practice these missions heavily so that everyone knows what an ideal performance looks like, practice until the mission can be executed reflexively and can be led by anyone in the group. At one time my outfit had one such "showcase" mission: we loaded up our P47s to the gills with bombs and rockets, moved out in immaculate formation and attacked as a body in a single run. The spectacle of 8-12 Thunderbolts hurtling down and letting off one hundred rockets simultaneously, and then dropping a barrage of bombs before streaking out in line abreast, was indeed something to be a part of, and something to behold. Such an attack may not cause massive destruction but it certainly raises morale tenfold!

20.2 COMMUNICATIONS

Air combat in general and formation tactics in specific requires crisp, clear and unfailing voice communications. In an environment where fights are literally won or lost in the bat of an eye you simply cannot rely on text commo. Aside from increasing the immersion effect voice comms increases unit efficacy by several orders of magnitude. However, for it to be useful, you need to enforce a strict radio transmission protocol. Voice communications must be brief, clear and to the point. With eight, twelve, sixteen pilots in the fight the chatter can become quite intense and it only takes one radio hog to mess it all up.

On the other hand, at certain times you can't talk too much on the radio. The "blower" is your primary means of maintaining control of your outfit and to keep up a wide-area SA, and when an element is engaged detailed comms may make the difference between death and victory. Thus, knowing when to speak and when to hold your tongue is critical.

In general, it's advisable to stick to the established format of RECIPIENT, SENDER, MESSAGE, however, if there's but you and your wingman or a flight of four where the pilots are intimately familiar with each other's voices then you can clearly dispense with such formalities.

Example transmissions using the standard format:
"Keyworth, this is Yardstick, follow me"
"Red Two from Red One, bracket"
"White Flight, this is White Leader, disengage north"

In combat, keeping your wingman appraised of your own action is paramount. He needs to know where you're at, where you're heading, what you intend to do – it's extremely easy to get lost and lose element integrity. E.g.
"Going in north, stay high"
"Off east, dragging"
"On your nine, clear"

The following table of radio transmission terms does not claim to be complete. It does however contain the most frequently used and useful transmissions and should suffice to establish a common language for your squadron, should you not already have one.

Radio communication	Meaning
[your, my] Six!	Enemy on [your, my] six
AA	Anti-aircraft (battery)
Bandit	Enemy aircraft
Bingo	Remaining fuel sufficient for RTB only
Blow through	Continue straight on after merge, do not turn
Bogey	Unidentified aircraft
Break	Break, enemy on your six
Buster	Best speed
Copy	Message received and understood
Clear	Your six is clear
Close up	Tighten formation
Cross turn	Execute level 180 degree turn towards wingman
Disengage	Break off combat in stated direction
Disregard	Carry on
Drag, dragging (direction)	Drag (am dragging) enemy on your six straight in the stated direction
EA	Enemy Aircraft
Extend (extending)	Open up separation to enemy
Free	I am not engaged
Friendly or blue	Friendly aircraft
Grab (grabbing)	Climb (am climbing)
In	Going in on enemy
In, in, in	All engage
Kick out	Increase formation width
No copy	Message not understood, please repeat
No joy	No visual (of enemy)
No tally	No visual contact
No threat	Enemy is not a threat, ignore
No vis	No visual contact
North, east, south, west – go!	Execute level 90 degree turn in specified direction
Off (Off off)	Breaking off attack (All break off)
On the runway	On the runway ready to take off
On the way	Heading to you
Out (heading)	Breaking off engagement (to regain energy)

Radio communication	Meaning
Pancake	Land
Reform	Expedite formation over given landmark or on given heading
Repeat last	Repeat last transmission
Reverse	Turn around (to engage)
Roger	Message received and understood
Roll	Take off
Stand by	Stand by (for further communications)
Tally	Visual contact gained
Unable	Unable to expedite or answer request
Up	Airborne
Winchester	Am down to last few rounds of ammunition

IN CLOSING

One pass, haul ass.

YOU and you alone are responsible for your six o'clock. Never expect a six call and you won't miss the one you never got.

Don't be shy to wallop the enemy with more than is required to down him.

If you take a hit, take it as a hint to return to base immediately.

Don't assume anything.

Greed kills.

Fly gently, kill swiftly.

Don't play with your food.

Live to fly, fly to live.

SELECT BIBLIOGRAPHY

Bekker, Cajus: The Luftwaffe War Diaries, Da Capo Press

Bond, Charles R & Anderson, T: A Flying Tiger's Diary, Texas A&M University Press

Boyington, Pappy: Baa Baa Black Sheep

Clausewitz, Carl von: On War, Princeton University Press

Demoulin, Charles: Firebirds!, Smithsonian Institute Press

Franks, Norman: The Greatest Air Battle, Grub Street

Frederick Blesse: No Guts, No Glory

Freeman, Roger A: Zemke's Wolfpack, Orion Books

Gabreski, Francis: Gabby, A Fighter Pilot's Life, Schiffer Military History

Galland, Adolf: The First and The Last, Buccaneer Books

Johnson, "Johnnie" : Wing Leader, Penguin Books

Johnson, Robert S: Thunderbolt!, The Honoribus Press

Lipfert, Helmut: The War Diary of Hauptmann Helmut Lipfert, Schiffer Military History

Lundstrom, John B: The First Team, Naval Institute Press

Milward, Alan S: The German Economy at War, London University Athlone Press
O'Leary, Michael: VIII Fighter Command At War, Osprey Aviation
Rudel, Hans-Ulrich: Stuka Pilot, The Noontide Press
Sampson, R.W.F: Spitfire Offensive, Grub Street
Shaw, Robert L: Fighter Combat, Tactics and Manoeuvring, Naval Institute Press
Smith, R T: Tale of a Tiger, Tiger Originals
Spick, Mike: Luftwaffe Fighter Aces, Greenhill Books
Spick, Mike: The Ace Factor, Naval Institute Press
Spurdle, Bob: The Blue Arena, Crecy books
Toliver and Constable: The Blond Knight of Germany, McGraw Hill

Made in the USA
Monee, IL
10 June 2026

52475357R00129